Of Kinkajous, Capybaras, Horned Beetles, Seladangs

Of Kinkajous, Capybaras, Horned Beetles, Seladangs,

ILLUSTRATIONS BY GLENN WOLFF

HarperPerennial
A Division of HarperCollins*Publishers*

AND
THE ODDEST AND MOST WONDERFUL MAMMALS, INSECTS, BIRDS, AND PLANTS OF OUR WORLD

Jeanne K. Hanson and Deane Morrison

A hardcover edition of this book was published in 1991 by HarperCollins Publishers.

First HarperPerennial edition published 1992.

Designed by Helene Berinsky

The Library of Congress has catalogued the hardcover edition as follows:

Hanson, Jeanne K.
Of kinkajous, capybaras, horned beetles, seladangs, and the oddest and most wonderful mammals, insects, birds, and plants of our world / Jeanne K. Hanson and Deane Morrison. — 1st ed.

p. cm.
ISBN 0-06-016464-6
1. Animals–Miscellanea. 2. Plants–Miscellanea. I. Morrison, Deane, 1950– II. Title.
QL50.H323 1991 89-46543
574–dc20

ISBN 0-06-092268-0 (pbk.)

92 93 94 95 96 CG/RRD 10 9 8 7 6 5 4 3 2 1

This book is dedicated to all the people I love.

—J.H.

To my parents, to Casey,
and to those who labor to save what is left.

—D.M.

Preface

Some will say, "This book tells me more about strange plants and animals than I really want to know." But keep it handy for the ideas it will give you. The future of the world depends on enough human beings realizing that we all need each other in strange and unexpected ways.

Along the Hudson River we have a campaign to clean up the waters. Because it is slowly but steadily succeeding, we have hope that this world may survive too, in spite of foolish, short-sighted people, and also in spite of those who can see far but have "tunnel vision"—that is, their knowledge is so specialized that they make foolish mistakes. Along the Hudson, when we sing "This Land Is Your Land," we add a new verse:

Woodland and grassland, and river shoreline
For every living thing, even snakes, bugs, and microbes
Fin, fur, and feather, we're all here together
This land was made for you and me.*

Keep learning, keep singing.

—Pete Seeger

Of Kinkajous, Capybaras, Horned Beetles, Seladangs

Smashing Crustaceans

Looking like a cross between a shrimp and a praying mantis, the stomatopods, also called mantis shrimp, inhabit burrows and crevices in warm, shallow seas around the world. (A few live in deep water, but little is known of them.) Though harmless looking, they pack a wallop far out of proportion to their half-inch to 13-inch size. Their strength lies in their two forelimbs, which are kept folded beneath their mouths, mantis-fashion. In the group known as spearers, the forelimbs are studded with sharp spines; when a shrimp, worm, or small fish happens by, the forelimbs shoot out and impale the victim. The prey is then drawn close and torn to shreds by the stomatopod's sharp mouthparts and associated appendages. Few animals can beat the sheer speed of a spearer's strike: It is over in four- to eight-thousandths of a second. And stomatopods have been clocked at more than ten meters per second. The speed is even more remarkable for occurring underwater.

In the other major group, the smashers, the forelimbs are heavily armored. Smashers can break open a clamshell with a

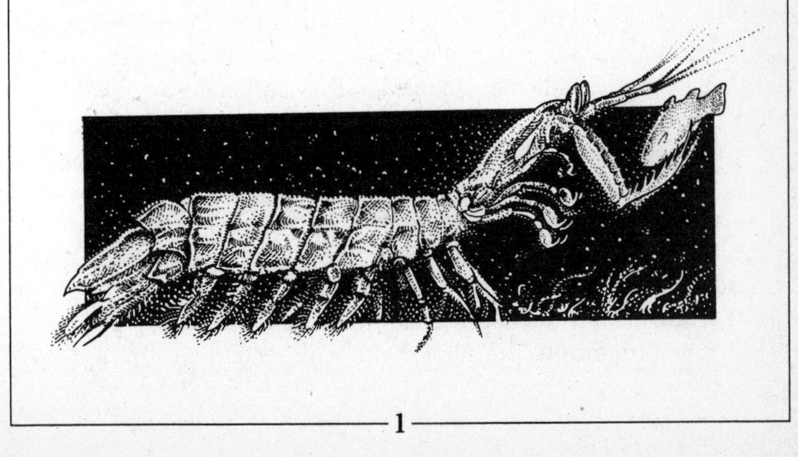

single blow, and one of the bigger specimens even broke an aquarium wall made of double-layered safety glass. But then, such a feat is little wonder for an animal that strikes with nearly the force of a small-caliber bullet. Not surprisingly, the formidable claws and carapace of a crab stand no chance against a smasher, which will batter a crab to pieces, then drag the crushed remains to its lair to eat them. Smashers can also spear soft-bodied prey by using the sharp tip of their forelimbs, which remain folded up during a smashing strike. For their relative size, stomatopods are stronger, faster, and fiercer than tigers. Unseen by most divers because of their reclusive habits, they nevertheless rank among the most aggressive animals in the world.

The Real Sea Monsters

Could sailors' reports of huge, sinuous sea serpents have any basis in reality? Or are they simply figments of the imagination, wild tales to entertain one's fellows in a tavern after a long and tedious sea voyage? Before dismissing these stories as tall tales, consider the oarfish. If any real-life creature resembles a sea serpent, it is this denizen of the deep. With its striking red crest and ribbonlike body, the oarfish is strange enough to give rise to a sea serpent legend. So unusual is its appearance, the real question may simply be whether sailors actually saw this elusive fish. For if they did, they could hardly have resisted giving it such a fanciful moniker.

The oarfish swims like an eel but has a much more elongate body. Its proportions are such that a ten-foot oarfish would measure only about five inches from top to bottom. Some grow to a whopping 35 or 40 feet or even bigger, as witnessed by a 56-foot "sea serpent" found in 1808 that was most likely a specimen. It is the longest known bony fish.

Adding to its formidable looks, a flame-red dorsal fin extends the entire length of its body like a flowing, fiery mane. The crest of dorsal fin rays on its head—as many as eighteen on a large individual—give the oarfish its Japanese name, which means "cock of the palace under the sea." And just as a cock can raise and lower its comb, so can an oarfish raise and lower its crest. Northern Europeans named it "king of the herring," because it was said to bode or accompany the beginning of a herring run. Indians of the Northwest Pacific coast regarded a relative of the oarfish as the "king of the salmon" for similar reasons. Killing the fish, the Indians believed, would mean the end of the salmon

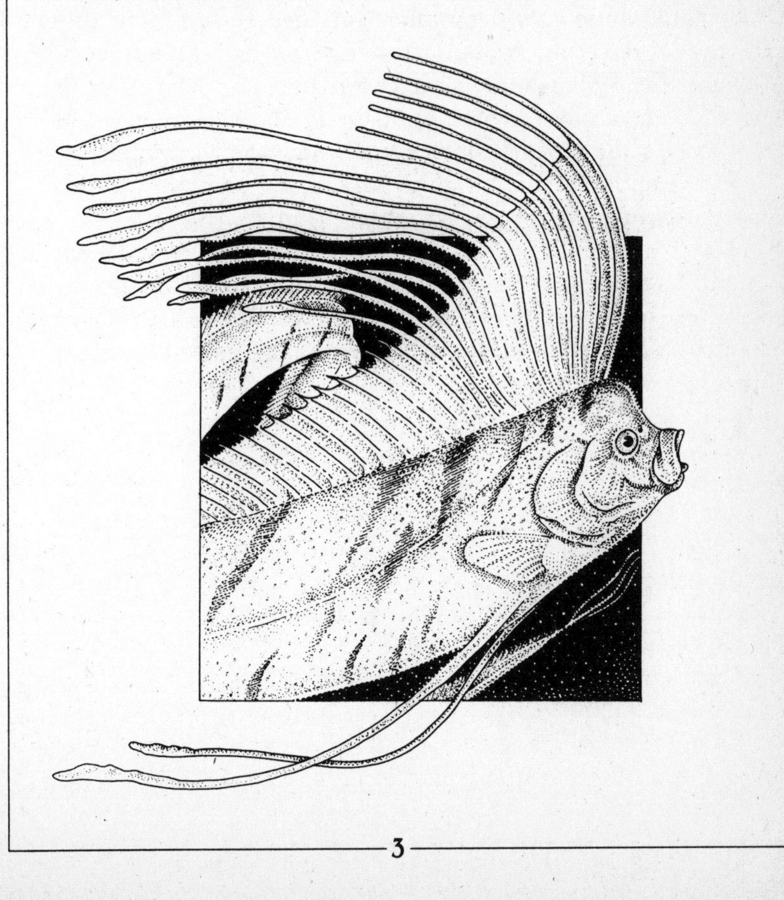

run. The fish's English name refers to its long and slender pelvic fins, which end in broad, bladelike tips resembling oars.

The oarfish's regenerative powers border on the miraculous. A male specimen chased ashore by a shark at Pompano Beach, Florida, in 1958 had been almost completely cut in three pieces, yet had healed and survived. Uninjured, the five-foot fish probably would have been ten feet long. In fact, almost every captured adult oarfish has borne signs of having lost and regenerated up to half its body. An oarfish's digestive system fills just under half the length of its body, and apparently a predator can bite off the tail up to that point without inflicting a fatal injury.

Despite its fearsome appearance and toughness, the oarfish poses no threat to humans. Adults live between about 300 and 3,000 feet underwater and feed on small shrimplike animals. The adult is the only deepwater bony fish known to be a filter feeder; it feeds by straining water through large numbers of long, spiny "gill rakers" as it passes over the gills. Even if it were to encounter a swimmer, its small, toothless mouth would be useless for biting. Adults usually manage to steer clear of oceanographers' nets, although larvae and juveniles are occasionally caught.

The sea offers few such striking contrasts between flamboyant looks and reclusive habits as this living legend. So next time you hear of sea serpents, think of the oarfish—a real member of the oceanic community whose truth may indeed be stranger than fiction.

In a Crocodile's Jaws

Many believe that the crocodilians—crocodiles, alligators, and their relatives—have but one use for their toothy jaws: to devour prey. But those mouths are made for more than biting. Crocodilians also possess strong voices, as the American alligator frequently demonstrates by "singing" distinctive "songs." Its relative, the caiman, a 6- to 12-foot denizen of South America and southern North America, can growl, snort, croak, snuffle, or, if wounded or seized by mating urges, roar like a lion. Many crocodilians, including the American alligator and the huge Nile crocodile, use their mouths to care for their young. For example, the female Nile crocodile buries her eggs in a riverbank nest and guards it for twelve or thirteen weeks. When the eggs are ready to hatch, she digs the nest open. As the hatchlings emerge from their eggshells, she picks them up in her mouth, packing in a clutch of eighty or more by depressing her tongue to form a kind of pouch. She releases them in shallow water by opening her mouth and swishing her head from side to side, and as the babies swim back to shore, they cleanse themselves of riverbank sand. The father sometimes takes eggs about to hatch into his mouth and rolls them gently between his tongue and palate until the hatchlings break free. He may also carry them to shallow water and release them as the female does.

Bifocal Fish

The four-eyed fish of South America has the unique gift of "dual vision." This finny submarine, complete with its own periscope, makes its home in the tidal fresh waters bathing the muddy mangrove swamps along the continent's northeast coast. The four-eye actually has two eyes, but each is split into a lower part for underwater vision and an upper part for air vision. The fish swims along near the surface, its upper eyes protruding from the water and leaving two V-shaped wakes that are its trademark. Every few seconds it dips below to wet its upper eyes, then bobs back and continues on its way. When spooked it dives a foot or two, but its buoyant swim bladder makes it hard to stay down. Soon it is back at the surface, going about its business and keeping an eye out for danger. Two related species share the bifocal trait; the most northern specimens are found in southern Mexico.

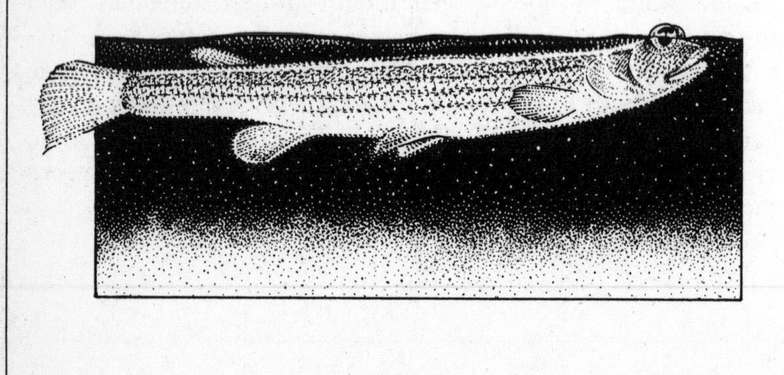

Men of War

Floating on the ocean surface, the Portuguese man-of-war looks like an innocent blue balloon, smooth but for a pinched-up crest on top. The innocence, however, is confined to appearance, for below the surface the man-of-war dangles death in the form of long, stinging tentacles. Named for the Portuguese warships of past centuries, the man-of-war is really a colony of tiny animals called coelenterates, relatives of hydras, jellyfish, and coral. Some members of the colony have the bell-shaped "medusa," others the "polypoid" form seen in coral and sea anemones. The entire mass is buoyed by a bluish gas-filled sac and subject to the whims of winds. Feeding is accomplished by polyps, from which extend tentacles bristling with nematocysts, the stinging cells that make encounters with jellyfish so unpleasant. Trailing up to 20 meters below and behind the man-of-war, the tentacles capture and sting to death fish and other prey, which are then hoisted to the mouths of the polyps. The tentacles, which rank among the most extensible of animal tissues, can be contracted to only a few centimeters in length. After feeding, the polyps pass the digested food to medusae through a network of cavities that serves as a gut. Indeed, the men-of-war, like the captains of their namesake, "run a tight ship." A human swimmer stung by a man-of-war may experience such extreme pain as to go into shock and drown.

Emperors of the Ice Continent

Standing as high as a tall man's hip, the emperor is the largest penguin and perhaps the hardiest. It lays its eggs on the Antarctic ice, and in winter, a time when temperatures can easily drop to minus 70 degrees Fahrenheit. It helps that emperors are very social and huddle together against the cold, but their feats of egg-hatching and chick-rearing still deserve admiration, if not awe. Most emperors breed on shelves

of ice extending out to sea. They flock to their rookeries in fall, and the females lay eggs from early May to June. The female, thin from weight lost during courtship and egg-laying, then returns to feed at sea, leaving the male to incubate the egg for sixty to seventy days. He holds the egg on top of his feet, close to his body, shielding it against the bitter cold and unrelieved dark of the Antarctic winter. If his mate hasn't returned by the time the chick hatches, he feeds it with a high-protein, high-calorie liquid secreted by his esophagus. If she has, she relieves him and gives the chick regurgitated food, perhaps fish, squid, or shrimp, as its first meal. Both parents take turns brooding and feeding the chick. For all the care, though, the life of an emperor chick is fraught with danger. Some 25 to 33 percent of eggs and chicks die, often from blizzards, undernourishment (if the weather is too bad for the parents to hunt successfully), or predation by birds, such as the giant fulmar, a type of petrel, or the skua. Chicks big enough to swim may die in the jaws of the leopard seal, a frequent predator of penguins. If they survive to adulthood, they will develop into accomplished swimmer-hunters capable of diving for more than 18 minutes and reaching depths of 250 meters. They will also inherit the empire of ice on which to raise their own chicks, continuing the fight for survival in what may be the most inhospitable nursery on earth.

Sex Changes in the Wild

The strange little spoon worm starts out as a larva floating in the ocean. Eventually it sinks, and where it lands determines its role in life. If it lands on the ocean floor, it becomes a female. If it lands on a female, it develops into a male and crawls into the female's uterus to live out its life. By no means unique, the spoon worm's sexual identity crisis is replayed in species all over the planet. Sex changes commonly occur in response to environmental change, often as an aid to reproduction but sometimes just to help organisms survive when things get though.

A crustacean that lives as a parasite on the heads of Caribbean fish changes sex to reproduce. The parasite, which may grow to three inches, is a male when it first lands on the fish's head. If a female is already attached, it stays male and fertilizes her eggs. If the female dies, it turns into a female and waits for another male to come along. Clown fish behave similarly. A large female lives with her smaller mate among the tentacles of a sea anemone, protected from their stings. When she dies, the male changes sex and takes a smaller young male as its new partner. Slipper-shell snails show a pattern reminiscent of the spoon worm. The tiny males come to rest on piles of the bigger females, where they father their offspring. But as they grow and as more males settle on top of them, they can reproduce more by turning female, and do. Prawns, a type of shrimp, may change to female after a year or two as male. The change occurs earlier in life when heavy fishing removes a large percentage of the females in a population.

The female stoplight parrotfish, a beaky denizen of coral reefs, has brilliant red and gray markings. But when she grows older, she turns into an equally brilliant-blue male. (Some males change from red to blue as they age, but without changing sex.) Another reef fish, the bluehead wrasse, also changes from female to male. On a small reef, large males keep harems of females and chase away any small males that try to intrude. So it is an advantage for a small fish to be a female and change to male only when it is large enough to keep a harem. On large reefs, the big males can't handle all the smaller males that try to "steal" the females, and such sex changes are less frequent. The blue-streak cleaner fish makes its living by eating parasites off other fish that share its tropical reef home. Cleaners live in schools composed of a single male and five or six females. The male is the largest, and he mates with all the females, who have their own social hierarchy. If a female dies, all of her subordinates move up a notch. If the male dies, the biggest female replaces him by turning into a male.

Sex reversals are the order of the day in marine fish, but rare in fresh-water fish. A fresh-water cichlid, however, was recently found to change from male to female and lay eggs in the absence of a mate. Perhaps it will be revealed that additional cichlids and other fresh-water fish have this ability as well.

In another recent discovery, African reed frogs were found to change sex. Females housed in predominantly female terrariums changed to males, probably, as in fish, to maximize their potential for breeding. Whether they do this in nature is yet to be determined, but it seems likely.

Animals are not alone in changing sex. Spinach plants, for example, can change gender when the soil gets dry. Dryness dramatically slashes the number of fruits each female plant can bear but reduces male pollen output much less. Thus male plants do better in dryness and females do better in moist conditions. The desert salt bush also can switch sex. Smaller female salt bushes may make only 100 fruits, compared to more than 100,000 for larger females, so male plants sometimes become female when they are big enough to take advantage of the chance to leave more offspring. Another plant, the jack-in-the-pulpit, first pushes through the forest soil as a little male plant. As a male, it makes pollen and so carries a lighter reproductive burden than females, which must produce seeds. After three to five years it takes on the duties of a female, but it may change back to male any time conditions become harsh and make it too difficult to provide nutrition for all of its seeds.

Sometimes the temperature determines which will be female and which male. American alligator eggs hatch into females alone if incubated below 86 degrees Fahrenheit and into males alone if the temperature is above 93. Many lizards, including leopard geckos and agamas, behave likewise. The Atlantic loggerhead turtle does it the other way around. Males hatch from eggs kept cool, females from warmed eggs. Oddest of all, the American snapping turtle hatches females from eggs kept at either temperature extreme, and males from eggs kept neither too warm nor too cool.

The Fastest Land Animal

Distinguished from leopards, the other famous spotted cats, by their lankier build, tufted shoulder hair, and distinctive black facial lines, cheetahs have been clocked at up to 70 miles per hour on short sprints. Usually, though, they pursue their prey at speeds closer to 40 miles per hour. If a cheetah gets close enough before breaking into pursuit, not even the swiftest gazelle can outrun it. Once downed, the victim is usually bitten in the throat and suffocated. Despite their size and hunting prowess, however, cheetahs defer to other predators out to steal their kills; lions easily chase them away from prey carcasses, and even an aggressive human may bully a cheetah into abandoning its prize. In other words, it pays more to be a cheater than a cheetah.

Cheetahs are now competing against people for something more important: survival. Already driven from India, the Middle East, and parts of Africa, cheetahs survive in protected areas of Africa. But even if all hunting and human encroachment on their habitat were halted, the outlook for the cats would still be clouded by the loss of genetic diversity in the species. A simple

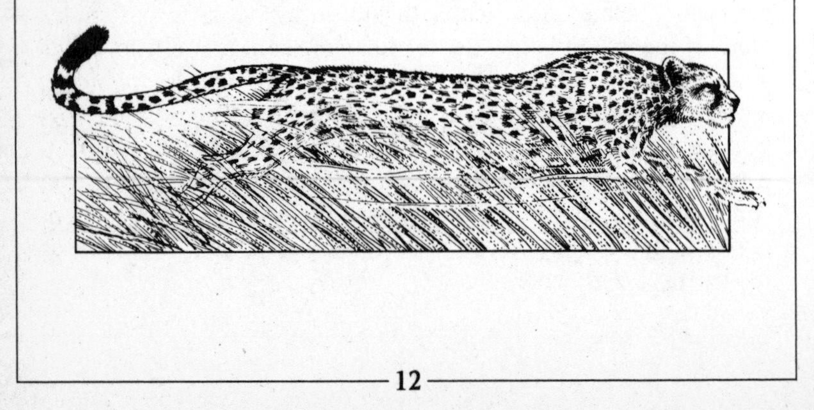

experiment illustrates the point. Scientists have grafted skin between unrelated cheetahs without the graft being rejected, evidence that the cheetahs' immune systems could not tell "self" from "stranger." That kind of interchangeability marks a highly inbred population, typically one that has passed through a "bottleneck"—a stage in which the breeding pool is so small that individuals end up mating with relatives. Even if such a population subsequently grows, the loss of traits once carried by nonrelated individuals persists, leaving a genetic make-up that offers too little variability to allow adaptation to new circumstances, such as climate change and a greater susceptibility to disease epidemics. All this adds up to a good chance that cheetahs will disappear from more of their range, but with strict protection they will have the best shot at long-term survival.

Tunneling Toads

The burrowing toad of Africa digs in for the dry season, diving headfirst into the soil, with its pointed nose leading the way. A mating male needn't bother to dig, however; he just hops on a female's back and rides piggyback as she digs down a few inches. Underground, they shed their eggs and sperm, and the female remains in the nest chamber to guard the developing eggs. When the tadpoles hatch, they wriggle down to the nearest water along an underground tunnel that their mother has built.

Walking on Water

Actually, quite a few birds *run* across water, but none is known to take a leisurely stroll without help from its wings. However, the jaçana, a robin-sized relative of the rails, comes close. Looking for food in its tropical marsh home, it walks over lily pads on feet with extremely elongated toes. The toes spread out the bird's weight and keep it from falling through the fragile lily pads in the same manner as snowshoes keep hikers from falling through snow. And Wilson's storm petrel, an oceanic bird, seems almost to walk on water as it hovers above the ocean and pats the surface with one or both feet. The patting stirs up small crustaceans and other food bits for the petrel, which is also about the size of a robin. When whole flocks flit over a patch of ocean and pat the surface, the petrels may look like a swarm of butterflies.

Water runners include ducks, geese, coots, loons, and many other water birds that rush at each other in aggression. The common loon, for example, goes after intruders with feet beating, wings spread, and bill open, a position called the vulture. Many waterfowl and diving birds must rush along the surface in order to become airborne. But the champion rusher is surely the Western grebe, a duck-sized water bird with a long neck, black feathers on top and white below, fiery red eyes, and a stilettolike beak, that winters in coastal areas of North America and breeds in interior wetlands. In one of the world's most spectacular courtship rituals, the male and female swim side by side, throwing their heads back and turning so that their bills point straight down, as if preening their back plumage. Suddenly, both birds rise up and dash side by side twenty to thirty feet across the surface, necks bent in an upside-down L and wings folded, their feet churning like propellors. Sometimes more than one competing male rushes along with the female. A pair seals their union after the rush when both dive and surface with a bit of weed in the bill. Circling each other breast-to-breast with feet pumping madly to keep their bodies out of the water, they point their bills upward and complete three or four turns before settling into their usual attitude—body mostly submerged and neck sticking up like a periscope.

In the insect world, water striders, whirligig beetles, and springtails are among those that can move across the fragile, elastic surface of very calm water. It is an ecological niche open to the truly lightweight and agile.

At the water's surface, the top water molecules are not pulled in all directions equally, as they are elsewhere within the water, but are stretched preferentially toward the water mass's center. This stretch creates a kind of membrane on top that can support an object—the insect—a bit heavier than itself. The phenomenon is known as surface tension. Even a few raindrops can disturb this delicate tension, and a film of ice, of course, destroys the balance; so under these circumstances, the insect heads for the edge of the land for safety.

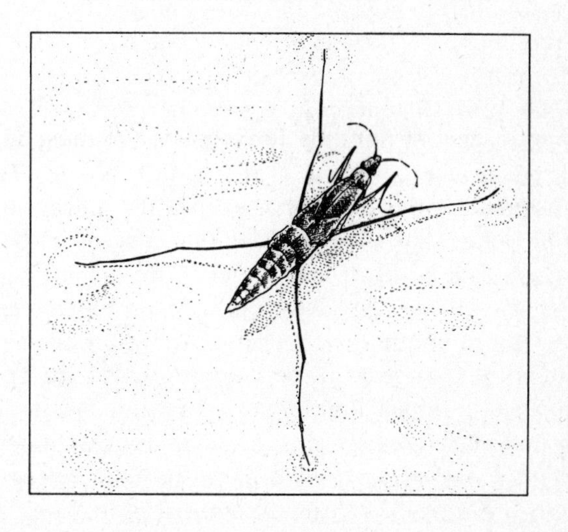

Under placid conditions these water creatures are quite acrobatic. Water striders, covered with stiff water-repellent hairs and scales, can stride, glide, and jump across the surface on only their feet and part of their hind legs. Whirligig beetles will swim in circles on the surface but can dive underwater, too. And springtails, the most ancient of these water Olympians, can spring forward by slapping their tails; always less than a third of an inch long, their entire bodies are waterproof, save for a tiny anchor organ that keeps them from blowing away in the wind. All of these insects, except springtails (which are scavengers), find their food by the tiny vibrations that their prey makes moving within the delicate surface film of their world.

Fantastic Sharks

Sharks account not only for some of the most dangerous fish, but also for some of the oddest looking. The hammerhead, for example, has eyes at the ends of two long, rectangular lobes that stick out at either side of its head. The goblin shark, a rare, deepwater variety with a sharp snout and protruding jaws, has a head that in side profile resembles an anvil. The megamouth shark, a large midwater species, has a mouth so huge it appears to be laughing when it's open. The mouth is also luminous (luminosity is not uncommon in deep-dwelling fish.) The cookiecutter, or cigar shark, a 20-inch denizen of deep water, glows a vivid green in its belly region. The eerie lights are thought to attract potential prey. The Port Jackson has a snout resembling a pig's (it is sometimes called the pigfish). Finally, consider the threshers, whose prodigious tails curve up and away from the body, accounting for half of the shark's total length. Threshers use their tails to stun prey fish, and some individuals have been seen using them to whack birds flying low over the water. With at least 350 known species, there's plenty of competition for the strangest member of the shark family.

The Hydra's Heads

At first mistaken for a plant, the humble hydra has amazed two centuries of scientists with its recuperative abilities. Cut it in half, slice it down the middle, turn it inside out—virtually no treatment is too rough for this sturdy little coelenterate, a relative of sea anemones, jellyfish, and coral. Less than half an inch long, the hydra's body is a tube only two cells thick, anchored by a "foot" at the bottom and crowned by a ring of tentacles around the top opening, which serves as a mouth. Many hydras are green and look like a flexible tree trunk, capped by the long, whiplike tentacles. Animals small enough to be caught in them are thrust down the inside of the tube and digested.

Many hydras appear to be either green or brown, a consequence of the food they store. Some grow living green algae in their cells and will turn colorless if kept in the dark, where the algae cannot live. Hydras move by a somersaultlike motion, in which they bend over and stand on their "heads," then bend again to touch down on their "feet." They have no eyes, yet are strongly attracted to light. Their nervous system consists of a diffuse "nerve net" of cells and fibers running throughout their bodies, with nothing resembling a brain. Hydras can sprout complete baby hydras from their sides in a process called budding.

Human manipulators have succeeded in turning a hydra inside out like a sleeve without causing apparent harm and have fused two individuals of the same species to form a new individual. When a hydra is cut in half horizontally, the foot end grows a new head of tentacles and vice versa, resulting in two new specimens. But they perform their most spectacular feat when the head is split in two: Each half-circle of tentacles grows to form a whole new head. If the process is repeated, a many-headed "monster" is created. This phenomenon suggested the hydra's name, which recalls the nine-headed serpent monster killed by Hercules, who found that when he cut off one head, two grew in its place, and which he finally dispatched by burning the severed stumps.

A Bear Built for Oblivion

Cute and docile, with big black eye patches set in white fur, the giant panda has become a Chinese national treasure. But does nature value the panda as much as people do? The answer seems to be no; while nature abounds with species obviously well adapted to their environment and able to take advantage of favorable environmental change, the panda appears just to scrape by. Only about 1,000 remain in the wild, fewer than 100 in captivity. With efforts to breed pandas in zoos still quite limited and Chinese panda refuges poorly protected from poachers and other human disturbance, little hope is held out that the big bear will be around fifty years from now.

The panda's difficulties are, in part, due to its reproduction, which progresses at a snail's pace. Females come into heat for only a few days each year, and if they give birth, they are likely to miss going into heat the next mating season. Pandas are very choosy about their mates, too, a trait that makes pairing them up in captivity especially chancy. In the wild, they eat bamboo almost exclusively but can extract only about 17 percent of its calories. Such an inability to obtain nutrition forces them to eat almost constantly and seems to prevent them from storing enough fat to hibernate like other cold-climate bears. This inefficiency means a new mother must forage for food while burdened with a naked, helpless cub instead of nursing it while she sleeps in a sheltered den. The difficulty of caring for the cub also means the panda mother, unlike grizzlies and black bears, can't raise more than one at a time. Indeed, pandas often give birth to twins but reject one cub. Because bamboo stays green year-round, it provides a steady food source; whole species of the giant grass, however, die off at regular intervals (usually between 40 and 120 years), leaving the pandas to search for alternates. This pursuit wasn't so difficult when the human population was relatively small, but now pandas are isolated in pockets and cannot easily move into a new range when food gets scarce. For example, a large die-off of bamboo in one area of China in the 1970s cut the local panda population in half. All of these factors lead inevitably to the conclusion that the world will lose its most popular bear unless ways are soon found to boost its breeding success.

Sibling Rivalry

Only about one out of three shark species lays eggs; the rest give birth to live young. Even in the womb, some of these sharks display a most sharklike appetite. Scientists were astounded to find embryonic sand tiger sharks with eggs in their stomachs, a sign that they had been eating their siblings-to-be. It turns out that the female sand tiger releases hundreds of pea-sized eggs during her pregnancy, which lasts for several months. In each of the two oviducts, one of the earliest eggs ovulated will develop into an embryo. The embryo then nourishes itself until birth by devouring subsequent eggs as the mother produces them. Nor are they fussy about food. One scientist was bitten by a developing baby sand tiger while feeling inside the mother. Sibling cannibalization has also been found in species including the porbeagle, crocodile, and thresher shark.

The Most Versatile Plant

Tall, flexible, and encased in a paper-smooth skin, bamboo has more uses than a Swiss army knife. It comprises a thousand species native to lush tropics and such temperate areas as Japan and the southeastern United States. Bamboo also shoots up faster than any other plant; one specimen was clocked at almost four feet in twenty-four hours.

Its size belies the fact that bamboo is not a tree but a grass. The largest bamboo makes a magnificent grass, indeed, with its foot-thick stem and 120-foot height. Unlike trees, most bamboo species are not solid but hollow, except for the horizontal nodes that give them their segmented appearance and much of their considerable strength. Also, bamboo doesn't taper toward the top like a tree but maintains a constant girth for its entire height. Bamboo grows from underground stems called rhizomes, offshoots of the parent plant. Much to the chagrin of bamboo farmers, however, each species reproduces sexually at regular intervals, usually 30, 60, or 120 years. When the breeding year arrives, all the bamboo plants of that species everywhere in the world produce flowers at once. Seeds develop from the flowers and fall heavily to the ground, and at that point the mature plants die. As early as the tenth century, Chinese historical records chronicled the simultaneous growth of young plants of the 120-year variety and, long after those who witnessed their sprouting had died, their blooming and death. Mass bamboo deaths have hurt not only farmers but Chinese panda bears that depend on bamboo for food. Recently, though, scientists have found a way to make bamboo flower in the laboratory, an advance that holds promise for producing seeds to replenish depleted forests.

Little of this botanical jack-of-all-trades gets wasted. The skin of some species makes a high-quality paper; the strong, flexible stems of others are fashioned into top-notch fly-fishing rods or sturdy scaffolding for buildings (some have stood up to typhoons that smashed steel frameworks); the tender young shoots are a delicacy in Oriental cuisine; and various parts of some species are used medicinally in treating ailments such as prickly heat, kidney trouble, and asthma. Some Asian peoples have even ascribed aphrodisiac qualities to fluids produced by certain bamboos. In addition, bamboo is a popular material for furniture, musical instruments, ornaments, and traditional weapons, such as swords or staffs. One more gift of bamboo must not be overlooked—beauty. Its smooth skin ranges from dark purplish-brown to yellow, green, and even gold. In some varieties, the nodes don't run straight across the stem, but at angles that produce intriguing geometric patterns. Such varieties are highly prized as ornamentals. Actually, bamboo need not dazzle the eye to be beautiful; given its contributions to human culture, bamboo is beautiful no matter what it looks like.

The Ship of the Desert

Bedouin nomads have good reason for calling the camel the ship of the desert. This living cargo vessel can move up to 600 pounds of freight during a long workday, and up to 1,000 pounds for short distances. During the cool winter it can subsist for three months without drinking, getting all the moisture it needs from the succulent plants in its diet. In the summer it can go at least ten days without water. In contrast to the relative barrenness of its surroundings, it provides peoples of the Near East, Middle East, and Africa with not only transportation but also nutritious vitamin-C-rich milk, wool, meat, leather, and dung that can be dried and burned for fuel.

The history of this remarkable beast of burden is closely tied to the history of the people who domesticated it 3,000 to 5,000 years ago. Without camels much commerce throughout the desert areas of northern Africa and southern Asia would not have been possible. The Moslem conquests of the Middle Ages were aided by the camel, which greatly surpassed the horse in maneuverability in treacherous sand or sandstorms. The fact that horses are easily spooked by camels didn't help Europeans in battle against Moslem camel corps, either.

Although found only in the Old World, camels come from a stock that originated in North America. Some of this stock moved into South America, giving rise to modern vicuñas, guanacos, llamas, and alpacas. Others migrated into Asia over a land bridge across the Bering Strait. Their descendants exist today as two species: the single-humped Arabian camel, or dromedary, and the double-humped Bactrian camel. The Bactrian, found in Central Asia, northern China, and Mongolia, is shorter and better suited to a cold climate. But it is the dromedary that has inspired awe, with its ability to survive in the desert heat and dryness.

For a long time people thought the camel stored water in its hump or elsewhere in its body, but the dromedary's hump is fat, not water, and no water reservoirs have been found in other organs. Even so, a camel can lose about a fourth of its body water and still retain 90 percent of its blood volume. Most of the lost water comes out of the tissues, leaving the blood intact. In contrast, a person who has lost a comparable amount of body water loses a third of his or her blood. What little blood remained would be thick and sluggish, and a strain on the heart. Under such conditions the heart would be unable to pump blood to the skin fast enough for heat loss to occur, and the person would soon overheat and die.

Have you ever had a fever of 105 degrees? Probably not; a human could not survive such a high body temperature for very long. But a healthy camel's temperature rises that high before

it begins to sweat freely. At night the camel's temperature may drop to only 93 degrees. Thus the animal must spend a fair amount of the day warming back up to 105, whereas a person would begin to sweat soon after the scorching sun rose in the morning. By tolerating a huge swing in its temperature, the camel greatly cuts the time that it spends fighting the sun. Besides being stingy with sweat, camels significantly reduce their urine volume when they are forced to do without water, again decreasing their body water loss.

The fat in the dromedary's hump serves another useful purpose. With most of its fat concentrated there, the camel, unlike most mammals, stores very little under the rest of its skin. This lack of fat limits the insulation and so helps the animal lose heat to the environment. The camel's back doesn't heat up, even though it sits beneath an insulating hump of fat directly under the broiling sun, because it is insulated by thick fur that even in summer grows several inches long. Heat from the 120-degree air flows toward the camel's skin more slowly through the dense wool than it would toward an unprotected human back. And don't forget, our skin is cooler than the camel's, which allows speedier heat conduction inward. To maintain our cooler skin, we must evaporate sufficient sweat, thereby increasing our water loss and compounding our difficulties of surviving in the desert.

Imagine that a camel has just made it through eight days of merciless sun with no water and finally arrives at its destination. The beast will look as though it hasn't eaten in weeks, with its ribs and hipbones showing through its skin. But not for long. When watered, the camel will replenish all of its water loss in just a matter of minutes, even if the loss is a quarter of its body total. An average camel can put away nearly 30 gallons of water in ten minutes. A severely dehydrated person would not be able to regain all the lost water at once; he or she would need to rehydrate over at least a few hours, as well as eat some food, in order to recover fully. Camels may have evolved the ability to do it so quickly because desert water holes are also frequented by predators, and so an animal that can shorten its stay will be less likely to encounter one.

Today camels are little used for transport, except by some of the poorer nomadic African tribes. Their days as the indispensable ships of the desert seem to have been ended by trucks and highways. Nevertheless, their adaptation to the desert remains a vivid example of how living beings fit into some of the most inhospitable environments on Earth.

An Egg-Laying Mammal

It has been said that the camel is an animal that appears to have been designed by a committee. If so, then the duck-billed platypus was designed by at least *two* committees—that didn't get along with each other. They produced a fascinating product, however. Native to streams and lagoons of Australia's eastern region, the platypus is definitely mammalian, yet retains several reptilian traits left over from the time when mammals evolved from reptiles. Its line may go back—independent of other mammals—150 million years.

The platypus looks like a duck, with its wide, flat bill and webbed feet. Its reproductive system is markedly reptilian, right down to the single orifice—called a cloaca—that it uses for both reproduction and excretion. Like birds and reptiles, it lays eggs. On the other hand, the platypus has a wide tail reminiscent of a beaver's, its body is covered by a thick brown coat of fur, and it nurses its young like any garden-variety mammal. Even its chromosomes are a mosaic, some of the reptilian type and some mammalian.

The platypus's bill is bony on the inside but soft to the touch. The seat of a sophisticated prey detection system, its surface is dotted with tiny pits containing organs that respond to the faint electrical signals given off by movements of crustaceans and other invertebrates. As the platypus swims along, eyes shut and bill wagging from side to side, it homes in on any signals it receives. At close range it snaps up the prey in its bill. Since it often hunts at night and in murky water, this system solves the problem of poor visibility. Prey are stuffed into cheek pouches for safekeeping while the platypus continues to hunt. After about ninety seconds of submersion it surfaces, empties the pouch contents into its mouth, and eats them. Juveniles chew with "baby teeth"—degenerate molars that disappear by weaning time—but adults grind their food with horny plates on their jaws. (The lack of harder teeth has obscured the platypus's evolutionary history, as teeth and jawbones are often the basis for identifying fossil mammals.)

The mostly solitary platypus starts looking for a mate around August. During the several days or few weeks of courtship, a pair may circle each other and nuzzle bills, then perform some splashy acrobatics together. In one peculiar stunt they swim in tandem with the male holding the female's tail in his mouth. Whatever they do, the female must be careful not to get in the way of the male's hind feet, each of which is equipped with a sharp, hollow spur containing a potent poison. The poison gland, located at the top of the leg, becomes active as mating season approaches. The spur helps its owner in fights with rival males and may help it acquire a territory. Spurring can be fatal to a platypus, and humans who have been spurred accidentally, or who have voluntarily taken a weak injection of the poison, report intense pain. After courtship and copulation, the female builds a nest of leaves and grass in her burrow. She usually lays two eggs and incubates them by holding them against her abdomen with her tail. The incubation period probably lasts ten or eleven days.

For all their small size—males reach only about twenty inches from snout to tip of tail, and females seventeen inches—platypuses do an outstanding job of keeping themselves warm, even during long winter hunting dives. Their fur traps a layer of air, which helps it maintain 30 to 40 percent of its insulative value in water, compared to only 4 to 10 percent for beaver or polar bear fur. Also, platypuses can more than triple their metabolic rate in response to cold water. Armed with such formidable adaptations, a platypus can maintain its body temperature for several hours even in freezing water.

Once hunted nearly to extinction for its soft, thick fur, the species is now protected. In response it has bounced back vigorously and today graces streams east of Australia's Great Dividing Range, from northern Queensland to the southern part of Tasmania. Undisturbed by people, the platypus may live and breed up to the age of twelve or so. Surely a survivor from the dawn of mammals deserves as much.

A Lot of Bull

Towering over most of its bovine cousins, the gaur of Southeast Asia is thought to be the world's largest wild ox. A healthy male may stand six feet at the shoulder, measure nine feet from head to rump, and weigh more than a ton—surely a lot of bull. An average cow would be a little smaller. Known in Malaysia as the seladang, the gaur roams the rocky, forested hills stretching from India to the Malay Peninsula. Though wild, it doesn't present much of a threat to humans unless wounded or spooked—definitely a mistake and a predicament to avoid. Distinguished by a large humplike ridge running halfway down its back, a dark coat with white stockings, and huge, deerlike ears, the powerful gaur can jump most cattle fences. Unfortunately, human expansion has helped its number dwindle to only a few thousand. But interest in the gaur is rising now that several U.S. research groups have begun

studying the potential of mixed gaur-domestic cattle breeds and have produced some hybrids. They hope to use artificial insemination and embryo transplantation to produce a steady supply of hybrids from parent species whose territories don't overlap. This could help especially in interbreeding domestic stock with another wild ox, the rare kouprey of Vietnam and Cambodia, which may be even less able than the gaur to spare individuals for transport to breeding farms outside the country. If these experiments succeed and the new mixed wild-domestic breeds catch on, we may find a different type of beef on our tables.

Forest Giants

Reaching toward the heavens like a living skyscraper, a 295-foot giant sequoia named General Sherman takes honors as the world's largest tree. Its weight, estimated at more than 6,000 tons, would balance more than 35 mature blue whales, the largest animals. Its base measures almost 110 feet in girth and 32 feet thick, and its largest limb is about 7 feet thick. General Sherman, estimated at 3,500 years old, contains enough wood for fifty six-room houses. It stands in Giant Forest, part of California's Sequoia National Park, along with other members of the redwood species known as giant sequoias. Also called the Sierra redwood, the giant sequoia thrives at altitudes of 4,500 to 8,000 feet, nourished by the moisture sweeping up the western slope of the Sierra Nevada from the Pacific. Its roots, though shallow, form an underground network three or four acres across. The tallest known tree, a 368-footer, belongs to the coast redwood species. Though taller than giant sequoias, coast redwoods weigh less and seem to sacrifice girth for height. The tallest redwood on record fits this pattern; its trunk is about 14 feet in diameter and 44 feet around. A third species, the dawn redwood, is native to China. A shrimp in comparison to its California cousins, it grows "only" about 100 feet tall.

Besides height, sequoias also rank high in toughness. Their two to four-foot-thick fibrous bark protects them from fire, and the large quantities of tannin in their wood discourages insects. In fact, no record exists of a mature redwood succumbing to disease or insects. However, they once fell in great numbers to the saw and are still prized by lumberjacks. Sequoias now enjoy protection, so we can continue to enjoy their breath-taking grandeur.

Insect Impersonators

If redwoods are the giants of the plant world, then surely orchids are the raving beauties. (At least, some male insects seem to think so, for they often try to mate with orchid blooms.) From the Arctic Circle to Tierra del Fuego, orchids flourish in marshes, woodlands, and even on other plants. After the spectacular eruption of the Indonesian volcano Krakatoa in 1883, orchids arrived with the first colonizers on the volcano's barren slopes, bringing to the ravaged land a speckling of color to relieve its moonscape appearance. These flowers aren't just pretty, they're tough.

Orchids display a staggering variety of shapes, sizes, and colors. The pink and yellow lady's slippers found in northern woods stand in stark contrast to the elaborate hanging strands of the yellow medusa orchid of Malaysia, named for the Greek mythological monster who had snakes for hair. The arms of the spider orchid of South America twist and curl like its namesake. The seeds of the white vanilla orchid, a tropical species, give the world one of its favorite flavors. But beneath the breathtaking beauty and variety of orchids lies one of nature's cleverest schemes for perpetuating species.

Many orchids fool male insects into helping the flowers "mate"—that is, cross-pollinate with other orchids of the same species—by mimicking the shape or smell of female insects. The males try to mate with the flowers and in the process pick up a sticky pollen body, which is carried to the next orchid that the male visits. One orchid looks just like a fly to the human eye and fools the male flies often enough that the insects, in trying to mate, brush up against the orchid's sticky surface and collect pollen. In the process they also deliver previously collected pollen to its stigma, thus pollinating the plant. The pouch orchids, including lady's slippers, trap insects and let them out through a "back door," where they pick up or deposit pollen. This is all fine for the orchid, but the hapless males waste their mating efforts, and sometimes even their sperm, on a futile quest. The insects' behavior is all the odder when one considers that "orchid" comes from the Greek word for testicle.

The flowers of one Ecuadoran orchid take a different tack and resemble male bees. The real male bees are extremely territorial and will attack intruders of their own kind. When a breeze sets one of these orchids waving, the local bee seems to mistake it for a rival and rams it, picking up a load of pollen. One orchid that doesn't bully its insect helpers, the yellow bucket orchid of Central and South America, obtains minerals from the nutrient-poor tropical soil via ants. The ants nest in the orchid's roots, and in turn they carry nutrients to the plant. The matching of orchid and insect is evidenced perhaps most spectacularly in the huge wax flower of Madagascar. The wax flower's pouch is more than a foot long and is pollinated by a moth that gathers nectar with a tongue of equal length.

Orchid fragrances may exhibit almost as wide a variety as their shapes. Each orchid's scent attracts a specific insect, perhaps a bee, wasp, beetle, butterfly, or, in the case of an orchid found in northern Minnesota bogs, mosquitoes. Orchid fragrances function as sex attractants, and insects often use them in this capacity, too. Some orchid fragrances smell exactly the same as mating attractants made by certain female insects. Scents are so individual that it may be possible to identify orchid species by their fragrances. Although the smell of a certain species may not be unique, it can stand out from others by varying its intensity or the timing of its release. Many orchids can turn their scents on and off. Some release their odor in response to light, while others respond to dark. Some turn it on at the same time of day without any obvious cue, with their own private clocks. One South American orchid showed just how precisely its fragrance release was controlled. A researcher wondered why he hadn't seen any pollinating insects on the orchid and so sat up with it at night. At 3:00 A.M. the orchid began to emit its scent. At 3:15 the man heard buzzing sounds, and at 3:30 the fragrance was gone. Drawing a flashlight, he examined the orchid and saw that it had been pollinated.

Orchids come in some 10,000 to 15,000 species, the greatest number of which grow in temperate regions. Curiously enough, their proliferation missed the Hawaiian Islands, since the islands formed too recently and too far from land for seeds to reach them by wind. But their wide distribution may be shrinking. The enormous popularity of orchids has led to an overpicking of the wild specimens and sales to nurseries that are often ill-equipped to duplicate the natural conditions that the orchids need to thrive. Several species are already threatened with extinction. So, the wise flower fancier will refrain from picking wild orchids and also from buying them, whether for cultivation or corsage, unless assured by a reliable source that the species is not endangered.

Living Fossils

What would you think if a young tyrannosaurus suddenly appeared on your doorstep?

A hoax, you'd say. Everybody knows that the tyrannosaurus died out in the late Cretaceous Period, more than 65 million years ago. Yet many animals and plants alive at the time of the fearsome "tyrant lizard" have survived to the present without obvious changes. Let's meet a few of them.

In 1938, fishermen off the coast of Madagascar netted a huge fish of a type they had never seen. It soon died, but they saved its remains and managed to get them to a biologist before they had decomposed too badly to identify. The biologist recognized the fish as a coelacanth (seal'-a-kanth), an ancient fish thought to have died out 90 million years ago. The discovery electrified the scientific world. In 1952 a second specimen was caught, followed by a few more. Unfortunately the coelacanths, accus-

tomed to the tremendous pressure at several hundred feet down, died within a day or two at the surface and so could not be studied close up. But the story isn't over yet.

Coelacanths belong to a fish group known as lobe fins, named for the fleshy protuberances at the base of the fins. Lobe fins first appeared about 390 million years ago, during the Devonian Period, the Age of Fishes. Their heyday came near the end of the Triassic, about 200 million years ago, followed by a sharp decline during the dinosaurs' reign in the Cretaceous. Coelacanths themselves had seemed to disappear 25 million years before the last dinosaurs.

In 1987, they again made headlines when a West German team descended into the western Indian Ocean off the Comoro Islands in a minisubmarine and photographed living specimens. The five-foot-long fish put on quite a show, doing headstands, swimming backwards and upside down, and moving some of its fins synchronously in a pattern common to four-legged creatures but rare in fish. Much remains to be learned about these living fossils, but research will be slow until scientists find a way to maintain them in captivity.

Meanwhile, Amazonian Indians have long considered another living fossil fish, the arapaima, quite a delicacy. With the biggest specimens variously reported at 250 to 400 pounds and 10 to 16 feet in length, the arapaima makes an excellent protein source. This largest of strictly fresh-water fish sports scaly armor and a primitive lung that helps it get by in the oxygen-poor waters it inhabits. Like the coelacanth, the arapaima is a relict of the Age of Dinosaurs. It seems to have survived fairly intact for 135 million years. But unlike the deep-dwelling coelacanth, the arapaima is a shallow-water fish, and so more vulnerable to human interference.

Several other fish also date back quite a while. A fresh-water dogfish, several species of garfish, sturgeon, and paddlefish are all throwbacks to the late Cretaceous Period, about 80 million years ago.

So is the Lanthanotus lizard of Borneo, whose ancestry comes close to that of the mosasaurs, the seagoing lizards of the Cretaceous that grew up to 50 feet long. The leatherback turtles found all over North America date to about the same time. These turtles are also called flapjack turtles because the lack of bone between the carapace (upper shell) and plastron (lower shell) makes them easy to cook. Even older is the very primitive, flat-headed South African clawed toad, a holdover from the Jurassic Period, about 180 million years ago. The tuatara lizard of New Zealand has been the only living representative of sphenodons, one of the four orders of reptiles, for the past 80 million years. Essentially unchanged for 200 million years, the tuatara can function in chilly weather that would immobilize other reptiles. Nevertheless, it takes life and reproduction at a leisurely pace, leaving itself vulnerable to the predatory rats infesting some islands.

Insects go back even further. Cockroaches first appeared in the Pennsylvanian Period, about 300 million years ago, and some termites lived in the Jurassic. Early beetles lived in the Permian Period, about 250 million years ago. Since then they have so multiplied that today about 60 percent of all animal species are beetles.

Such success stories notwithstanding, the real Methuselahs of the animal world are the little "lamp shells," or brachiopods, which have survived for 500 million years. These ocean dwellers, only an inch or two long, are often mistaken for bivalve mollusks, like clams or mussels. The mistake is understandable, since lamp shells also have a double shell. But while bivalves enclose the left and right sides of their occupants, the brachiopod shells enclose the top and bottom.

The 300-million-year-old horseshoe crab looks and acts like a regular crab but is really a relative of the spider. Crowned by its brown, domed, horseshoe-shaped carapace, it moves sluggishly along the ocean bottom on five pairs of walking legs, bringing to mind the fossil trilobites of the Paleozoic Era. People who encounter one should beware of a bite from the chelicerae, or pincers, that it uses to crush worms and other food.

Australia boasts perhaps the most astounding and delightful tale of animals with what might be called fossil habits. Both the duck-billed platypus and the spiny anteater, or echidna, still lay shelled eggs, something all other living mammals scrapped about 100 million years ago.

From Africa comes the okapi, a denizen of the equatorial rain forests of Zaire that looks like a much shorter version of the giraffe, to which it is related. It goes back about 30 million years.

The plant world has it own tenacious examples of such survival. Town dwellers may find the gingko tree, the only remnant of an order dating from the early Triassic Period, about 230 million years ago, along a residential street. Native to China and Japan, this lovely tree, with its uniquely shaped leaves, has been identified in fossils from Europe, North and South America, and Asia. It is planted in American cities, where it is valued for its resistance to smoke, cold, and dryness. And last but not least, the magnificent magnolia also has a long history. Magnolias date from the early Cretaceous and are typical of a group that gave rise to all flowering plants. Most Americans associate the tree with mint juleps and gracious Southern living, but it most likely began its reign as food for dinosaurs like triceratops.

Tadpole Meringue

A family of gray African tree frogs, called Rhacophorines, puts on a spectacular mating display. The female exudes not only eggs but a copious foam as she mates on a branch overhanging water. The male uses leg movements to spread the eggs evenly throughout the foam. Other males may join in, creating a huge, slippery mess of foam and frogs. The next morning, the foam hangs from the branch like a dollop of cooked meringue, its outer layer hardened to support the young and keep moisture sealed in. When the tadpoles are ready to hatch, the tough shell dissolves, and they drop into the water and swim away.

How to Choke a Tree

True to their name, the strangler figs of Old and New World rain forests choke the life out of the trees that give them their first foothold in the dense growth. Related to the edible fig, a strangler begins innocently enough when a bird or fruit bat drops a seed onto a large tree. Soon it produces a plant with roots that quickly head downward. Some roots snake down the tree trunk; others, called aerial roots, grow straight down through the air. Aerial roots may dangle for 90 feet or more before reaching ground, where they branch to form a stable base. Once in the soil, the strangler's roots tap into the ready supply of nutrients and moisture, and its crown really starts to expand and compete with the host tree for sunlight. Meanwhile, the trunk roots keep growing down, thickening and interlacing

to form a tight network around the trunk. Eventually, this corset of roots may choke the host tree. A tree so killed will rot away, leaving the circle of strangler roots to form its own hollow, interwoven "trunk." Some strangler figs, like the banyan of Southeast Asia, grow to gargantuan size. One Indian banyan's crown measured a whopping 2,000 feet in circumference, and other banyans shade whole Indian marketplaces. One Sri Lankan banyan shades an entire village. Strangler figs don't always kill their host trees, but they do hurt them by stealing sunlight. Nor are they the only strangler plants. They share the habit with others, such as the pink- or white-flowered clusias of tropical American forests.

Do Piranhas Really Eat People?

The scene has become clichéd: A jungle explorer, clad in khakis and pith helmet, wades into a river holding a gun over his head. The next thing you know, the water turns bloody as he screams and thrashes in mortal agony. His buddies at the camp rush to the scene to find only his clothes and pith helmet floating. "A gun's no use when the piranhas get you," the head of the expedition remarks soberly.

Movies and adventure comics have painted the piranha as a bloodthirsty killer, keen to attack any person or animal that falls into the water. Given this backdrop, it may be surprising to learn that the piranha, called *caribe* in Spanish, can actually behave tamely. Twenty species are known, but only five eat meat. Still, its menacing reputation is not entirely undeserved; it probably grew out of a few actual events that mushroomed into legend. The lesson seems to be that while bloodcurdling encounters with piranhas are rare, we still mustn't forget the damage that they can do.

Part of the peril stems from their habit of swimming in schools, where they pose the greatest threat, especially in muddy waters. They stage their famous frenzies mostly when the water is low and food is scarce. They pose less of a danger in full, clear-water streams. Piranhas frequent the rivers of Colombia, Venezuela, the Guianas, Brazil, Paraguay, and the areas in south to central Argentina, preying on small fish but also taking dead or injured animals when lured by the smell of blood. With their sharp teeth and blinding speed, they leave victims, once targeted, little chance to escape. One species grows to two feet—indeed the stuff of Hollywood horror fantasies—but most are much smaller. Their bodies are high and flattened from side to side, with a fleshy adipose fin between the dorsal and tail fin.

Indians of the Orinoco and Amazon basins eat piranhas, and some use the three-edged teeth to fashion arrowheads for small game. The jaws also make good scissors to cut different types

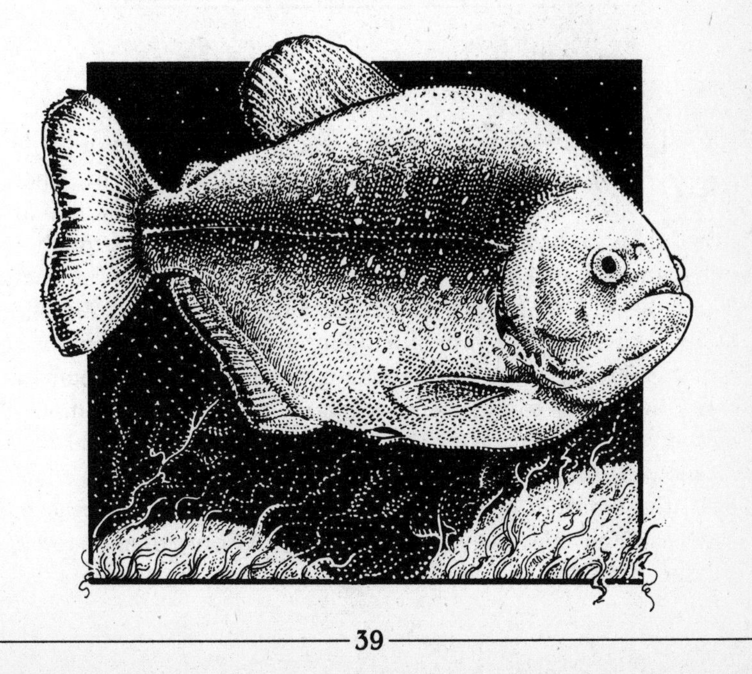

of foods. When worked by themselves, the jaws make a clacking sound like castanets. To attract piranhas, the Indians bat the water, because the fish think a disturbance means food. Unavoidably, fishermen sometimes suffer a good nip to the finger or toe when pulling piranhas out of the water.

Do piranhas always attack human bathers? No. People can swim in piranha-infested streams without being eaten alive. In fact, one film crew donned wet suits to get a fish-eye view of piranhas and swam among them for a long time, sustaining only one minor, almost accidental, bite. There is no reason why people and piranhas can't get along; after all, we kill more of them than they ever will of us. We just eat *them* fried instead of alive.

Four Stomachs in One

Like all ruminants—the grazing animal families of cattle, sheep, and goats—cows have four-part stomachs. Their elaborate digestive systems, which constitute a full quarter of their body weight and hold about nine times as much as our own stomachs, allow them to eat food very rough in fiber. The first and by far the largest section, the rumen, and the second, or reticulum, contain billions of microorganisms that break down the cow's food, including cellulose, which humans can't digest. These chambers regularly regurgitate soft masses of coarse food (the cud) for the cow to rechew. When these are swallowed, they pass from the reticulum into the other chambers, the omasum and the abomasum, and finally into the intestines. The cow digests not only the energy-rich materials produced by the microbes, but the microbes themselves.

Blood, Sweat, and Tears

Don't look now, but lots of innocent-looking insects are out to drink our bodily fluids. Forget mosquitoes—start thinking about moths that suck blood and tears, and bees that seem to have a thing about sweat. These creatures are startling illustrations that no group of animals, or plants, for that matter, is beyond evolving a strange new twist to its expected way of life.

Add to your list of bloodsuckers the Southeast Asian "vampire" moths. These moths are a far cry from the lepidopteran standard in both diet and dining utensils. Most moths feed by unrolling a long, tubelike proboscis into a flower's hidden pool of nectar and sucking up the sweet liquid. The proboscis is the moth's portable drinking straw and little else; it isn't a cutting tool and cannot pierce human skin. The vampire moth, however, can drive its sharp proboscis through skin and suck human blood for 10 to 60 minutes at a sitting. When it departs, it leaves behind a wound that continues to bleed for a few minutes and itches for considerably longer. This is a radical departure from the familiar clothes moth, which does no harm except to munch on clothes in its larval stage.

What made the ancestors of vampire moths stray so far from the beaten path in their feeding habits? One scenario goes like this: In the past, competition for nectar may have become so severe that some moths turned to drinking the juices of overripe fruit to survive. To do so, they would have needed mouthparts that could break the skin of the fruits. Such fructivorous moths, relatives of the vampires, do, in fact, exist in Southeast Asia. Some can bite effortlessly through peach skin, and one species can even pierce an orange. As long as fruit was plentiful, this strategy would have served the vampires' ancestors as well as it serves their modern cousins. If the supply were cut short, say, by a loss of certain seasonal foods, the ancestral moths may have turned to a more reliable source of nutrient fluid: mammals. Available year-round, they offered a steady diet for any moth that could evolve a proboscis and a metabolism to handle the rigors of bloodsucking. In this manner, using fruit as a stepping stone, some moths may have journeyed from blossoms to blood.

In the same area of the world lives another moth, specializing in tears. It may have evolved from bloodsuckers that once feasted on the leftovers of mosquitoes but found a steadier source of nourishment in the host animal's eye secretions. And why not? Tears are a fountain of nutrition, containing not only salt but proteins, white blood cells, and epithelial cells. These moths have been seen drinking from the eyes of deer, pigs, antelope, horses, buffalo, and elephants. Given a chance, they will drink even from human eyes. An adventurous scientist who described the moths' habits and allowed them to drink from his own eyes (after letting vampire moths suck his blood) noted how the irritation stimulated the flow of tears. For thirty minutes he put up with the discomfort, but how long most animals do is another matter.

Across the Pacific, in the tropical South American forest, live several species of stingless bees. (Other species live in the Old World tropics, but not nearly as many as in South America.) Some get almost as big as regular honeybees, but other species are twelfth-of-an-inch midgets. Some species seem to be attracted to sweat, while others make straight for the hair of the head, where large numbers can become entangled. In either case, they are capable of causing endless annoyance. Though incapable of stinging, they can bite, and this does nothing to add to their popularity.

Are Sloths Slothful?

The Spaniards who explored the New World were as free with their sarcasm as with their militarism. Encountering an animal that hung in trees all day, scarcely budging even when threatened, they named it the Nimble Peter. English speakers did little better, forcing the harmless creature to share its name with one of the seven deadly sins. In defense of the explorers, those names did capture the essence of the sloth; its whole life is geared to inactivity, right down to the algae and insects that make their home in its fur. Now, after centuries of insults, it is reaping well-deserved praise for its ability to survive life in the slow lane—a strategy no other mammal has pursued nearly as successfully.

Central and South America are home to two species of sloths: the two-toed and the three-toed, so-called for the number of toes on their forefeet. (Both have three toes on their hind feet.) Four times as common as its two-toed cousin, the three-toed weighs only nine pounds, compared to the two-toed's fifteen. The two-toed eats leaves, flowers, and fruit and keeps nocturnal hours. The three-toed subsists on leaves alone and is active—sort of—both night and day.

"Hanging around" is easy for both sloths, thanks in part to the long, curving claws that can hook over tree branches. In addition, sloth limbs are built to withstand long periods of hanging upside down, which allows a sloth to use less energy hanging than most other animals use standing up. Sloths spend so much time in this position that their hair grows from the belly toward the back, giving an upright sloth the appearance of having had quite a fright. So securely does the sloth hang that a dead one can hardly be pulled down but must first be raised to unhook its claws.

Both species frequent the high canopy of rain forests, where they blend in so well with the foliage that they are nearly impossible to spot. Early censuses undercounted sloths and reported that they lived only in cecropia trees. As it turns out, sloths are very common, accounting for about 70 percent of the total weight of arboreal mammals in many rain forests. They may outnumber the common howler monkey by ten to one. The reason that they seemed to be found only in cecropia trees is that the cecropia, with its spare, not-very-bushy form, was one of the few trees in which they weren't hidden from view. Sloths actually frequent several dozen tree species. Their treetop camouflage comes partly from the green algae that grow in their fur and partly from their slowness, for often it is movement that gives away an animal's presence.

How slow is a sloth? It may take half a minute to move a leg a few inches. A mother rushing full speed to help her threatened infant was clocked at 14 feet per minute. The algae growing in the sloth's coat help support the "sloth moth," which lives in its host's fur, along with a variety of beetles, mites, and ticks, all undisturbed by any efforts on the sloth's part to dislodge them. Sloths can't even sneeze fast.

They digest their food slowly too. Their disproportionately large stomachs act like fermentation chambers in which bacteria break down the leaves to yield energy. The three-toed passes food so slowly that it needs to defecate only about once a week. The sloth also saves energy by dropping its body temperature as much as 12 degrees Fahrenheit every night, something not many animals do without hibernating. In the morning, the sloth suns itself in the treetops to warm up again.

The three-toed sloth's diet makes it impossible to keep in captivity. Before the reason was discovered, some zoos tried to keep the animals but found that they would starve to death even with a stomach crammed with leaves. The explanation is that different sloths eat different leaves. The list of trees one sloth dines on may not even overlap with a neighboring sloth's list. Each sloth acquires its list as an infant, when it licks fragments of leaves from its mother's mouth and picks up both a taste for those leaves and the specific gut bacteria to digest them. With perhaps 200 tree species to choose from, each sloth feeds on

about 40, so competition between sloths for food is low. Unfortunately for the sloths that died in captivity, the narrowness of their diet made it easy for zookeepers to choose the wrong leaves to feed them.

The sloth diet doesn't allow much body building. Sloths are little but skin and bones, with only about half as much muscle as comparably sized mammals. It's just as well, since maintaining a large muscle mass requires a lot of energy, and energy is one thing sloths don't have in abundance. Their physique comes in handy, however, when they must cross a river or lake to get to a new feeding ground. Aided by their big stomachs and low muscle mass, sloths float like balloons, helped along by some of the world's most lethargic swimming movements.

None of this means sloths are passive. Though three-toeds rarely lose their temper, two-toeds have shown signs of irritability, including the will to take a vicious swipe at an intruder. In addition, sloths have been reported to survive mutilations that would kill almost every other vertebrate, and they can survive thirty minutes underwater. The latter ability is probably due partly to their slow metabolic rate, which keeps their need for oxygen low. Also, their upside-down position makes it difficult for big cats to grab them. In fact, a wakeful sloth would most likely spot a predator first, since its neck vertebrae allow it to turn its head a long way around. This also makes it easy to feed without moving.

Sloths may not be the brightest of mammals, but they manage a life span of at least fifteen years and sometimes twice that. They are the only living relatives of megatherium, an extinct, elephant-sized ground sloth frequently mentioned in books about dinosaurs and early mammals. Perhaps one day they will go the way of megatherium, but for now they don't seem to be in much of a rush.

Spitting Snakes

Depicted among the hieroglyphs of ancient Egypt are two spitting cobras representing Upper and Lower Egypt. This iconography has a basis in nature. When a threatening animal gets too close for comfort, the cobra does, in fact, shoot its venom into the victim's eyes, causing intense pain and at least temporary blindness. The cobra's aim is devastatingly accurate at ranges of up to about six feet. It manages its feat with the help of fangs that point up and outward instead of down. If spitting doesn't deter the visitor, the snake will resort to biting.

According to one report, a dog hit by spitting cobra venom scratched its eyes, allowing the poison to enter its bloodstream, and died. Later, a friend of the dog's owner was also hit in the eyes and felt such intense pain that he had to be restrained from scratching his eyes, even though he was an experienced animal collector and well aware of the consequences. Temporarily blinded, he slowly regained his sight.

Sociable Felines

From the house tabby to the mighty tiger, most cats naturally tend toward solitary living. Lions, however, live in groups called prides. Adult females form the core of the pride, with cubs and adult males coming and going. Adult females tend to be related to one another, and even take turns nursing one another's cubs. Strong adult males four years of age or older will join a pride for a couple of years before being deposed by other males. Those few years give the females time to raise the cubs that they have with each partner, an important point, because new males successfully invading a pride will kill any cubs already there before siring new litters. They will also chase away any subadult males, a tactic that removes them as competitors for food.

Females, being leaner and lighter than males, are better hunters and make most of the kills, but adult males quickly move in and claim the "lion's share" of the carcass before allowing females and cubs a turn. Lions do most of their hunting at night or just before dawn. Despite their ferocity, they fail to bring down prey more often than they succeed. Sometimes their efforts lead to their own death, as when a sharp-hooved zebra, in a last desperate attempt at escape, delivers a jaw-breaking kick to the pursuing lion. Unable to eat, the lion soon dies. Lions also scavenge much of their food and may appropriate the kills of smaller predators, such as cheetahs and hyenas.

Why do lions socialize? Lions, unlike tigers, live in areas where both predator and prey exist at high densities. Females are likely to encounter strange males, and large groups of females are better able to protect their cubs from would-be invaders. Although pride dwellers must share prey, this doesn't diminish an individual's lot, because the prey are large. With abundant, wary prey and much competition from other lions, it makes sense for lions to band together to defend hunting territories and cubs.

Survival of the Ferret

Its light fur broken up by black "stockings" and facial markings, the black-footed ferret ranks among the most appealing members of the weasel family. Such cuteness may have fanned scientists' eagerness to help the ferret, once thought extinct throughout its range, which extended from southern Canada, through about ten American states, and down to the Mexican border. Rumors of extinction proved exaggerated in 1981, when a ferret colony popped up in Wyoming. The colony grew to more than 100 individuals, but in 1985 plague struck prairie dogs, the ferrets' main food, followed by a distemper outbreak in the ferrets themselves. Reduced to only 17 survivors, the colony faced a bleak future. In saving them, wildlife scientists performed a minor miracle.

The scientists took the little band into captivity. Protected from disease and predators, the survivors thrived. But the pampered ferrets had to be taught how to fend for themselves, including the crucial skills of hunting and, of all things, mating. Mating lessons were rather tricky. The young males raised in enclosures didn't know how to mate, so the scientists brought in females from a related but less aggressive, partly domesticated, species and let them teach the black-footed males the facts of life. The wildlife experts also plan to train females to hunt prairie dogs, then breed them and let them raise their young in a larger enclosure, with prairie dogs. The mothers, they hope, will then teach the little ferrets to hunt.

The training paid off in a colony that grew from 18 to 118 ferrets between 1986 and 1989 and seems well on its way to a population of 500 by 1991, when the first are scheduled to be released back into the wild. Meanwhile, some individuals have been moved to start new colonies at zoos in Virginia and Nebraska, a step that will prevent a single disaster from wiping out all the survivors so carefully raised. If all goes well, the black-footed ferret's masked visage will soon peer out once again from its native prairie grass, signaling a big victory for a small but scrappy fighter.

A Cell as Big as a Doormat

On your next walk through a forest, keep an eye out for rotting logs and tree stumps. If you find one, carefully turn it over, or break it open. Inside the hollow of the stump or under the log you may find a large, flat, brightly colored splotch of what looks like thick mucus. Look closely and you may see that the mass is very slowly flowing over the wood. If it is, chances are you have just found a slime mold.

Slime molds belong to an unusual group of organisms called protoctists, which occupy their own kingdom in the biological scheme of things (the other kingdoms are plants, animals, fungi, and bacteria). At some points in their life cycles they are indistinguishable from the amoeba, while at others they behave just like regular, card-carrying fungi. Biologists once classified them with the fungi, perhaps because their reproductive structures are so funguslike, and gave them a Greek name meaning "slime fungus." But it is their amoebalike movements that make them most fascinating to scientist and hobbyist alike. Two varieties exist: plasmodial and nonplasmodial. The big, creeping ones, which may be colorful or colorless, are plasmodial slime molds. In contrast, nonplasmodial slime molds are all but impossible to find because of their tiny size. Let's turn the spotlight on plasmodials first.

The most spectacular phase of this slime mold's life cycle appears in the sprawling sheet of colored protoplasm one might find under a log. Called a plasmodium, it may measure up to a meter long but comprises only one cell. The cell is full of nuclei generated by one nucleus that divided many times without any accompanying division of the rest of the cell.

The amoebalike plasmodia send out "feet" of protoplasm, first in one direction, then another, like a flat mass of warm jello sloshing very slowly back and forth. Some spread out in a lacy, fanlike network of spidery projections. In this way they crawl over logs, engulfing and feeding on bacteria. The microscope reveals streams of protoplasm flowing along channels inside the cell in the direction of movement. How a plasmodium manages to move this way without benefit of bones or muscles has intrigued scientists for many years. The motion resembles not only that of amoebae but also that of white blood cells, and so some medical scientists trying to unravel the secrets of how our own white cells move within our bodies have turned to slime molds for insight.

When feeding conditions become poor, a plasmodium starts acting like a fungus. It produces spores in structures called fruiting bodies, which come in a striking array of colors and shapes, depending on the species. Some fruiting bodies have been likened to brown hair growing on trees; others form big black globs or pink puffballs. Inside the fruiting bodies, the single-celled spores mature. When conditions are right, the spores are released and scattered by the wind. After settling, they become amoebae, and in some species develop flagella for propulsion. Sometimes two amoebae join sexually to form a new amoeba,

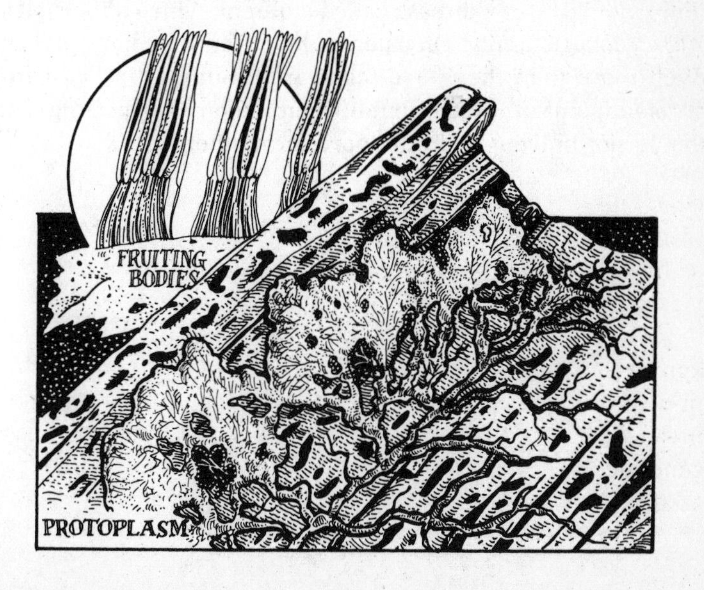

but in many species the amoebae remain independent. In either case, a single amoeba begins to multiply its nucleus and develops into a plasmodium, completing the life cycle.

Under adverse conditions a plasmodium may not form fruiting bodies but instead develop into a hard, dry, stationary mass called a sclerotium, which can survive long periods of inactivity without drying out. When conditions improve, the sclerotium becomes a plasmodium again.

The nonplasmodial slime molds also spend much of their time as amoebae. But instead of dividing to form a multinucleate mass, they congregate to form sluglike creatures a few millimeters long. Because the slug is such a composite, these slime molds have also been dubbed social amoebae. The amoebae find each other by means of a chemical trail laid down in their slime. In the laboratory, amoebae congregate from all over a petri dish—a distance of a few inches—to form a slug. However, no one is sure how far apart amoebae can be in forest soil and still follow a slime trail to the other amoebae with which they will form a slug. These slime molds also produce fruiting bodies, tiny structures about as tall as a dime is thick.

Of the approximately 500 species of slime molds living in forests around the world, only a few make trouble. For example, a slime mold causes a disease called clubroot, which afflicts plants of the cabbage family. On balance, however, they have probably given more than they have taken from humans, at least from the standpoint of wonder and intrigue. Their unappealing name should not prevent us from appreciating them.

The Barking Pika

Scampering busily around the rocky crevices of its mountain home, the round-eared, eight-inch pika looks like a rodent but actually belongs to the lagomorphs, the same group as rabbits and hares. Pikas, also called mouse hares or rock rabbits, eke out their living in the harsh environment above or near the timber line of mountains in western North America and northern Asia. Both North American species live in colonies beneath rocks, as do most of the 12 Asian species, while a few Asian species prefer to burrow. Even when hidden, they reveal themselves by their habit of "barking" vociferously at each other. Covered everywhere—even on the soles of their feet—by short fur colored from red to gray, pikas neither sleep through the winter nor store fat under their skin. To survive four months of snow, they need a healthy underground larder. Each pika provides one for itself by cutting and sun-drying several heaps of vegetation to form hay piles, then storing them in its own winter quarters. Unlike the nuts that squirrels bury, the pika's plant foods generally contain few calories, so great quantities must be gathered. Pikas vigorously defend their caches against thieving members of their own species, although they rarely, if ever, inflict serious wounds. Working hard at their task during the warmest parts of the day, these industrious little animals certainly know how to "make hay while the sun shines."

The Grinning Turtle

Camouflaged by its splotchy green-brown skin and a coat of algae on its shell, the matamata turtle wiggles its ear flaps and other appendages on its head, waiting quietly for fish to be attracted to its living lures. When one gets close enough, it suddenly opens its mouth and widens its neck, creating a partial vacuum that helps suck in the fish. The matamata, a denizen of northeastern South American rivers, has a neck almost as long as its shell and nostrils that stick up from its face like a double-barreled shotgun. Stretching neck and nose, the matamata can breathe underwater like a snorkler. When threatened, it pulls in its head as any other turtle would but curls it under its shell. With its wide, V-shaped mouth, the matamata appears to grin slyly, as well it might if it had any idea how strange it appears to humans.

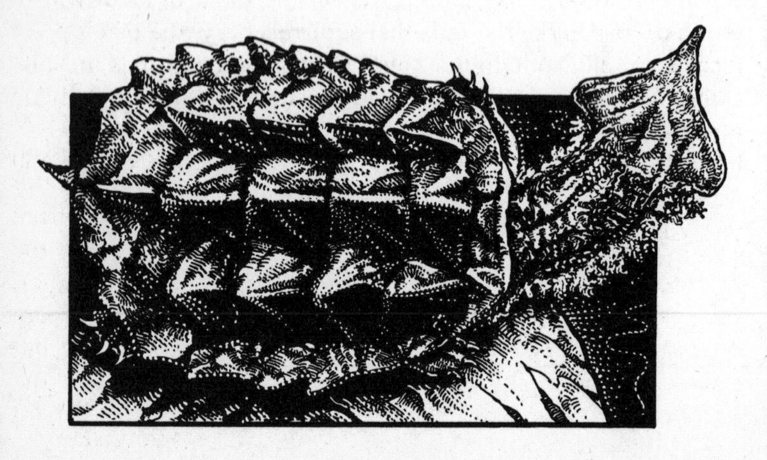

Spawning on the Beach

The beaches of southern and Baja California host a spectacle that might be called the greatest show on birth: the annual spawning rites of the grunion. This small fish, which grows up to about seven inches, lives in the near-shore waters but comes up on the sand to breed. Spawning occurs between late February and early September, but only at high tide on three or four nights after a new or full moon. Riding the tidal waves up onto the sand at the peak of high tide, the fish spawn as the tide begins to ebb. Each female burrows into the sand and releases her eggs, while males coiled around her shed their milt. The fish then flip back into the water. The eggs, sheltered from tides in their sandy cradles, develop in about a week. But they do not hatch until two weeks have passed, when the new moon becomes full, or vice versa and a second very high tide occurs. It appears that the agitation of the tidal surges are necessary to bring about hatching. Once hatched, the young are swept out to sea by the tide.

Far to the north, the capelin, an oceanic relative of the smelt, performs a similar spawning ritual on beaches in a circumpolar band including Iceland, Japan, and Newfoundland. Studies in Newfoundland indicate that eggs hatch two to four weeks after the June spawning, and larvae remain in the beach sand from hours to days before heading out to sea en masse. Unlike the grunion, the capelin larvae pay no heed to time or tide. Rather, their highly unpredictable departures happen when onshore breezes bring warm surface water, full of planktonic food but relatively few jellyfish and other predators of capelin, to the beach.

Wingless Flight

True wingless flight seems beyond any animal except the helicopter pilot, but expert gliders like flying fish and flying squirrels practice the next best thing. The most skilled species of flying fish inhabit tropical seas, though tropical fresh-water species live in Africa and South America. They achieve flight thanks to huge winglike pectoral fins and, in some groups, similarly enlarged ventral fins that act like a second set of wings. The California flying fish, a big species with four "wings," starts its flight by swimming just under the surface, its upper tail lobe often in the air. Spreading its pectorals, it takes to the air, except for its lower tail lobe, which it beats vigorously to gain speed. Finally it opens its ventral fins and takes off in a flight that may continue for 1,000 feet, though usually more like 100 to 300 feet. In most cases the fish rises 4 or 5 feet above the water; the record for flying fish lies between 25 and 36 feet. Flights generally last ten seconds or less, with at least one timed at forty-two seconds. Flying fish can speed along at about 35 miles per hour, and no, they don't flap their fins. Some land with a graceful dive, some with an awkward splash. They most likely fly to escape from predatory fish, but sometimes, attracted to on-board lights, they come aboard boats.

Floating through the air with the greatest of ease, flying squirrels travel from tree to tree or from tree to ground with the help of a wide flap of skin extending along either side from the wrist of the foreleg to the ankle of the hind leg. Close observations of North American squirrels indicate that they gain stability from their flattened tails, and in landing, exercise pinpoint control by lifting or lowering their legs to hit exactly the right spot. Right before touchdown they jerk up their heads and tails to slow down, then land vertically, hind feet first. Some of the longer flights carry the nine-inch rodents about 50 yards. Their relatives include five species of giant flying squirrels, which range in eastern parts of the Old World. The largest get up to three or four feet from nose to tip of tail and weigh nearly three pounds. They have been seen gliding 450 yards, banking during flight, and, apparently, ascending air currents up deep valleys. The gliding opossums of Australia and neighboring islands (not to be confused with the opossum of North and South America, which they superficially resemble) float to the ground in much the same way as flying squirrels. The best gliders travel more than 100 yards.

Even snakes get into the act. The golden tree snake of Indonesia offers the best example. This serpentine glider flattens its body and then holds itself rigid to achieve aerodynamic stability. A tree dweller, it moves among branches by jump glides and can proceed right to the ground in a single leap from a very tall tree. Its acrobatics are matched by its fierce bites, so it would be quite unpleasant to be at the receiving end of one of its graceful glides.

A Life without Water

At first glance, it seems unimaginable that any mammal could live without water. Even a camel would never make it in a dry, barren expanse of desert where rain came infrequently, water holes were nonexistent, and dew did not form every night. Yet certain mammals do live in these desolate places. Some, like the pack rat, subsist on succulent plants, such as cacti. But more impressive are the desert rats, which get by even in the scorching sands of Death Valley on a diet of dry seeds and no water at all.

Desert rats comprise many species scattered throughout the arid regions of Australia, Africa, Asia, and North America. A typical example is the kangaroo rat, native to the deserts of the southwestern United States. Kangaroo rats are not related to the kangaroo but resemble them in their long tails and strong hind legs, which they use to hop along the ground. They are 12 to 15 inches long, with the tail accounting for more than half that length in some species. The tail is tufted at the end and is used for balance and as a rudder when the animal turns at high speed. Their heads and eyes are quite large and mouselike, complete with cheek pouches for food storage.

Solitary, nocturnal creatures, these kangaroo rats spend days in their private burrows and venture forth at night in search of seeds, grass, and tubers. Sometimes they hoard their food in shallow holes, retrieving it after it has dried and storing it in their burrows. Since they eat dry or almost dry food, they can obtain moisture only by oxidizing hydrogen in their food to form water. Hydrogen is contained in all three classes of food—fats, proteins, and carbohydrates—but fat has more than the others and so yields more water when it is metabolized for energy. Protein has about the same amount as carbohydrates, but much is lost along with nitrogen in urea, a breakdown product of protein. Because urea requires water for its excretion, an animal cannot eat too much protein if its water intake is limited. So the rats eat seeds and other plant foods to provide them with enough protein for life, but not too much more, along with plenty of fat.

The kangaroo rat could never live on the amount of water it extracts from food if it did not also restrict water losses from urine, feces, and evaporation. Its water losses from urine stay low, thanks to powerful kidneys that excrete urea using only one-quarter the amount of water that a human kidney would need to excrete the same amount. In addition, the kangaroo rat can pack salt into its urine until it is twice as salty as sea water. In fact, kangaroo rats forced to drink by being put on a high-protein diet are able to drink sea water without becoming ill. The rat's feces are also very dry.

The third route of water loss, evaporation, gets a little trickier in the kangaroo rat. Like all other rodents, it has sweat glands only on its toe pads, but fewer of them. It appears to lose very little water through sweat and very little from the skin, where sweat glands are lacking, but there remains the problem of what to do about losses from the lungs and respiratory tract. After all, a mammal can't just stop breathing because the air is dry. But the kangaroo rat avoids the worst strain on its water supply by staying in its burrow during the hottest part of the day. Moreover, the air in its burrow is considerably more humid than the air in the desert above, further reducing respiratory loss. Without the extra humidity the rat would not quite be able to conserve enough water, and it would either perish or have to find a new environment in which to live. That it doesn't have to is a tribute to the effectiveness of its burrowing habit and one of the most efficient water-conserving physiologies ever discovered, which together make the kangaroo rat the dean of desert dwellers.

The Deadliest Sharks

Of the approximately 350 known shark species, 3 have been blamed for the large majority of attacks on humans. Leading the pack is the great white, subject of the film *Jaws,* followed by the tiger and bull sharks.

The largest great whites grow to at least 26 feet and can swallow an adult human, whole. They seem particularly prone to striking in Australian waters, whose popularity with bathers make Down Under a hot spot for shark attacks. Also known by such monikers as maneater and white death, the great white sometimes attacks boats and keeps at it until they sink. The tiger shark, recognizable by the vertical stripes along its dorsal half, grows to at least 24 feet. The bull shark terrorizes not only the open seas, but some fresh-water areas as well. It has been found 2,300 miles from the ocean in the Amazon River, in Lake Nicaragua, the Zambezi River, the Ganges River, where it has attacked pilgrims bathing in the holy waters, and as far upstream as Alton, Illinois, in the Mississippi River.

Most shark attacks occur in warm water, where bathers are more likely to be found. Many attacks may simply be cases of mistaken identity, in which the shark mistakes a bather for food. For example, a skin diver and a seal look a lot alike in profile, and a surfer paddling a small surfboard with head, hands, and feet dangling out might resemble a giant sea turtle as seen from below. Also, some species, including the great white, seem attracted to the color yellow. While no reliable chemical deterrent has been found, the U.S. Navy has developed an inflatable, floatable bag called the Shark Screen, which looks like a dull, shapeless bulk from below and works well to shield its occupant from attack. Precautions against shark attack include not swimming alone, at night, at dusk, near a fishing area, with an open wound, or during a menstrual period; leaving the water if fish seem to be acting strangely; not wearing bright objects, such as jewelry; and avoiding deep channels or shallow areas near a rapid drop-off. In an unfamiliar area, ask local people if any swimming spots have been frequented by sharks—and if so, swim somewhere else.

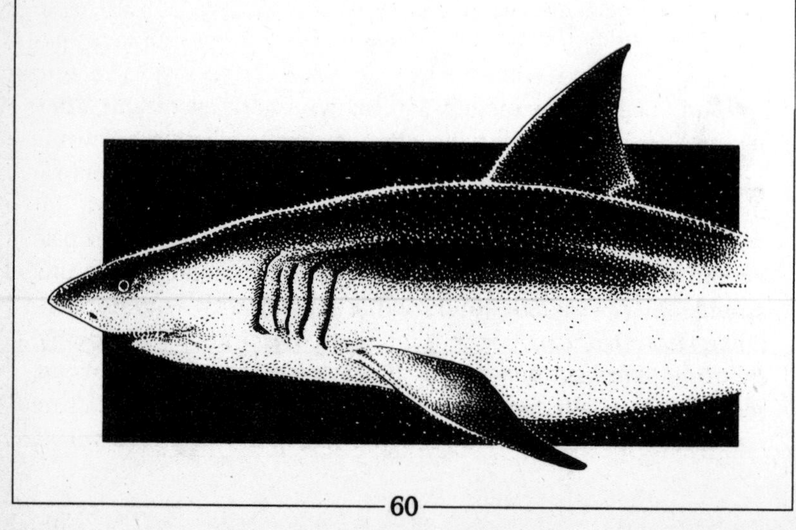

The Face That Launched a Thousand Charter Flights

Cute, round, and cuddly as a teddy bear, the gray-furred koala has become Australia's biggest tourist attraction. This sleepy-eyed little mammal has probably drawn more visitors to the Land Down Under than any other animal, including the kangaroo. In return, Australians are now trying to preserve the koala's future. If nothing is done, Australia's unofficial mascot could fall victim to the twin threats of epidemic and habitat destruction.

Like most other Australian mammals, koalas are marsupials and carry their young in pouches. Their resemblance to bears is just a coincidence, but they do have one thing in common: a heavy sleep schedule. But while bears pass the winter months in hibernation, koalas spend eighteen hours a day asleep in the nook of a tree. They are normally nocturnal, which must make life difficult for captive koalas when the tourists come to pet and fondle them in the middle of the day. Occasionally, a cute little fur ball will nip a person who gets overly familiar. They can also give a person a nasty scratch with their sharp claws.

The claws are just one of many features that make the koala a natural at tree living. Koalas are clumsy on the ground, but their superb balance and tailor-made hands and feet add up to arboreal artistry. Their hands are equipped with an opposable thumb and first digit, good for grasping a tree trunk or branch. Their feet also sport what might be considered an opposable "thumb," too, and the first two toes are fused. Thus the koala gets rock-solid attachment and support from its feet while using its hands to feed. In addition, koala rumps are equipped with a

flesh-and-fur pad that does quite nicely for sitting and snoozing over long stretches.

Anyone who has seen koalas only in the arms of an adoring keeper would be surprised to see them in the wild. Researchers trying to capture koalas have been greeted with screeching, belligerent subjects that don't take kindly to being prodded and dropped into nets. And during mating season, the males set up territories, fight with each other, and may chase after an unwilling female with young. They are, after all, wild animals, and simply because we perceive them as lovable does not mean that they always oblige.

Koalas eat only eucalyptus leaves, which are eaten by other arboreal marsupials, but are toxic to many animals. Koalas broke through the tree's chemical defense by evolving a way to detoxify the oils in the leaves that cause the problem. About 35 of the 600 or so eucalyptus species have found a place on the koala menu. Such a leafy diet means a heavy intake of cellulose, which the koala cannot digest; its intestines, however, are equipped with a well-developed fermentation chamber, where microorganisms do the job for it. Thus nourished, koalas typically grow to about twenty pounds and live up to 12 years in the wild, 20 or so in captivity.

Little koalas get their start between early spring and midsummer, the koala mating season. Females generally mate and produce one cub every other year; gestation lasts about 35 days. At birth, the cub is only three-quarters of an inch long and weighs less than one-quarter ounce, yet it must climb unaided from its mother's vaginal opening to the pouch where it will complete its early development. But climb it does, into the pouch—which is rare in that it faces to the rear—and fastens itself onto a teat. After seven months in the pouch, the cub takes up residence on its mother's back. At about one year of age it is on its own. If food is abundant, it will grow to nearly two feet.

No predators disturb the wild koala, but lately an eye disease has been taking its toll in blindness, pneumonia, and death. A program of capturing wild koalas and treating those with the disease has had some success, and a vaccine may be available in a few years. But with surveys showing up to 45 percent of the population affected with some form of the disease, koala helpers have their work cut out for them.

Koalas now enjoy strict protection, but it was not always so. Aborigines hunted them for food for thousands of years, and when Europeans arrived in Australia, the koala experienced the usual fate of the fur bearer: wholesale slaughter. The pelts were shipped to Alaska, where they wound up as fur hats. In 1927, just before koala hunting was banned, 600,000 pelts were exported from Queensland alone. Today, that number could hardly be harvested even if all koalas were killed. Wildlife experts estimate that only 10,000 to 50,000 wild koalas remain, and they are threatened by the logging of eucalyptus forests. This destruction of habitat not only deprives them of food and shelter but crowds them together and facilitates the spread of disease. That includes an eye disease, which is spread by sexual contact. Koalas have already died out in the southwest corner of Australia, a site of heavy urban growth.

On the bright side, many people have awakened to the threat and taken action. Tree-planting campaigns have helped, notably in the New South Wales community of Port Macquarie, where eucalyptus replenishment and civic awareness have helped human and koala to thrive side-by-side, even in the urban sections. Also, Australia is home to about 30 privately run reserves where koalas are kept safe, healthy, and in the public eye—and reach—a fortuitous link between preserver and preservee that can only help sustain conservation efforts. Given the nearly intractable problems of human expansion and the loss of wildlife habitat, the winsome koala needs all the good will that it can muster.

A Frog in the Throat

Getting rid of a frog in your throat is, for us, a mere annoyance. But for two related species of Australian amphibians, it's a necessity. In a bizarre incubation rite, these mother frogs swallow their eggs and brood them in their stomachs, releasing them (through the throat, of course) as little froglets. In one species the mother's stomach stops producing hydrochloric acid and loses some of the roughness that characterizes a normal stomach lining, but in the other the lining appears just like a regular stomach. In a similar scheme, the male Darwin's frog stores his offspring in a huge vocal sac. He, too, spits out his progeny as miniature froglets.

Are Earthworms Good for Anything Besides Fish Bait?

Burrowing earthworms make Swiss cheese out of topsoil, but it's all to the good. Their tunnels keep the soil loose so that plants can root deeper and oxygen can reach down to the root tips. Loose soil also holds water better, making it available to plant roots and reducing the likelihood of erosion. In addition, the worms act as garbage disposals, eating any organic food that they come upon, including feces, and turning it into a high-quality fertilizer. The worms produce feces called casts, which contain lots of nitrogen, phosphorus, and potassium. Earthworms themselves make nutritional snacks. A robin tugging a reluctant worm out of its hole is about to enjoy a 60- to 70-percent protein meal. A smattering of tribal peoples have traditionally partaken of earthworms, but worm meat doesn't seem to have a future in many cuisines. It may supplement livestock and poultry feed, especially in Southeast Asia, but that's about it.

Approximately 3,000 species of earthworm populate the world's soils. They need a damp environment under about 82 degrees Fahrenheit in order to thrive, and soil provides a good escape from heat and dryness. The worms produce slime to coat their tunnels and further reduce evaporation. When it gets too damp, say, after a heavy rain, they wriggle to the surface to keep from drowning. The biggest earthworm, an Australian variety, grows to 12 feet and makes gurgling sounds like water being sucked down a bathtub drain. Encountering one would be quite an experience on a dark and stormy night.

The Tiniest Deer

The deer family ranges in size from the mighty North American moose to the diminutive South American pudu. But while the moose is relatively familiar, few people have seen the reclusive pudu. Only 12 to 15 inches at the shoulder and 15 to 22 pounds, the two species of pudu both have short, spiky antlers and hardly any tail. About as big as terriers, they would be easy to miss in the woodlands that they inhabit. The northern species lives in the Andes of northern Peru, Ecuador, and Colombia. Its coat ranges from reddish brown to dark brown or gray. The southern, native to southern Chile and southwestern Argentina, usually wears a rich brown coat. When disturbed, it takes refuge in water, although it doesn't swim especially well. Both pudus prefer living in small groups. Fawns complete their growth in only about three months and are ready to mate in a year. Its rapid growth, however, has not protected the pudu from the threat of extinction. Hounded by

hunters, its forest habitat ravaged by burning, the tiny deer is now being bred in nine zoos worldwide, in hopes that a healthy population can be built up for eventual return to the wild.

Where Birds Meet to Mate

If a male is trying to attract a female, the last thing he wants is another male around, right? So why do some male birds flout conventional wisdom by hanging around with buddies at breeding time?

Whatever the reason, the male buddy system, known to biologists as lekking, appears in species with short-term or no pair bonds, including some insects, fish, and mammals, as well as birds. Bird lekkers include sharp-tailed and sage grouse, hummingbirds, tropical jewel birds, birds of paradise, two species of prairie chicken, and some shore birds. The hangout is called a lek (from either a Swedish or Old English word for "play"), and it can be as varied as open sagebrush country, where fifty male grouse congregate, or a twig in a dense forest where two or three jewel-bird swains gather. In any case, the males wait at the lek for females and display vigorously to them when they arrive. In a large lek, a relatively few dominant males do most of the mating, and many males don't mate at all. In a small lek, only one male may do all the mating.

Presumably the nonmating birds allow this apparent privilege because they derive an advantage from it. Even without an explanation, lekking behavior makes riveting bird-watching. The sage grouse and the jewel bird lek in completely different ways, so perhaps a look at them will best indicate the boundaries of this peculiar behavior.

The male grouse begin to congregate in February, when warm days first return to the windswept sagebrush plateaus of the Rocky Mountains. Through March, April, and early May, the males come early to the lek and spend three or four hours "strutting"—cocking their tail feathers, ruffling their neck feathers, and making a popping noise by expelling air from esophageal sacs. Each of the 50 or 60 birds has his own small territory within the lek and sticks to it throughout the season. Females arrive at the lek in flocks during about three weeks in April, congregating near the middle of the lek, where the most dominant males have staked claims. Usually, each female mates with one male, with males in the middle being heavily favored. Most males, in fact, go unchosen; only 10 percent or so of the males do about 90 percent of the mating. Afterward, the females leave to build their nests and raise six to eight young. The nest may be kilometers away, and the females will have nothing more to do with males until the following year.

The females don't come to the lek randomly. Older, more experienced birds generally come earlier and lay their eggs earlier in the season than their novice companions. The older ones probably guide the younger ones to the lek and "show them the ropes." Interactions among females seem to have much to do with their subsequent mating behavior with males. And while the males sometimes have disagreements over territorial boundaries, they rarely, if ever, hurt each other.

The beautiful red-, black-, and blue-plumed male manakins, or jewel birds, found from northwestern Costa Rica to southern Paraguay, prefer to congregate on tree branches with only one or two comrades. Females attracted to the lek by the males' calls witness a bizarre courting ritual, in which the males cooperate with each other as they display to the females, although only one male, the dominant one, will do all the mating. The willingness of a subordinate male or males to pitch in and help the dominant male corner the market makes this behavior one of the most highly evolved bird behaviors known.

The jewel-bird partnership starts out when a male lays claim to a 10,000-square-foot "court" of forest. In the case of one species, the long-tailed manakin, he will be joined by just one other male. The partners select a main perch from which to call for females. If a female arrives and settles on a perch near the main one, the males sit side-by-side on the main perch facing her and take turns leaping into the air and fluttering their wings, emitting coarse, low-pitched calls as they jump. If she chooses the main perch, the males line up in front of her, one behind the other. The first in line performs his leaping display, then flies to the back of the line and lets the other male take his turn. This freewheeling pattern may be repeated 100 times or more, speeding up as it goes. Meanwhile, the female watches every move intently, hopping up and down and beating her wings with excitement. The display ends when the dominant male utters a high-pitched note as he takes his final turn. He then circles the female in a slow flight and mates with her.

Male swallowtail manakins form partnerships of four to six, with each male holding a rank in the hierarchy. The top bird usually does all the calling, and when a female arrives, the display is very similar to the long-tails' freewheeling. Again, the dominant male gets to mate, most if not all of the time.

Whether a species forms leks or not seems to depend on its diet. Research on New World birds, including manakins, indicates that species that don't lek, but instead form monogamous pair bonds, tend to feed on insects, whereas lekking species favor fruits. Apparently, fruit is plentiful and easy to obtain, so males of fructivorous species can afford to spend time and energy maintaining a lek. Insects are harder to find, however, and insectivorous males and females must form partnerships to ensure enough food for both a brooding parent and offspring.

Diet seems to work differently in Old World birds of paradise. Monogamous species tend to favor fruits such as figs, which are easy to consume but are available at unpredictable times, compelling the birds to hunt constantly for them. The fruits are also low in certain essential nutrients. It is thus advantageous for a male to pair-bond with a female and help her search for food for their young. Lekking species will eat figs when they're available, but tend to favor insects and fruits such as nutmeg, which are rich in fat and protein and available throughout the year. The predictability of their food supply enables females to feed themselves and their nestlings and frees males to spend time in leks. Research also indicates that some male birds of paradise take more than one mate, but don't form leks. In these species, females tend to have relatively small, nonoverlapping ranges, so not many are likely to visit any single lek site. Only when females' ranges overlap considerably is it worth it for males to lek.

No one can say why subordinate males participate in leks, but scientists speculate that they gain something from their "apprenticeship." They may learn better display techniques from the dominant males, or they may increase their chances of moving up to dominance by staying put, since the second-most dominant bird will normally move into the top spot if it is vacated. Subordinates would probably fare worse by leaving to set up a new court, since resident birds have an advantage over interlopers when rank is in dispute. A more general theory of animal behavior holds that animals may act selflessly for the benefit of relatives. In the case of subordinate male birds, their behavior could promote the breeding success of related males and, therefore, foster the survival of some of their own genes. Not enough is known yet about lekking birds to say which theory might explain the phenomenon. Whatever is going on, lekking remains one of the animal world's premier spectator sports.

Playing Possum

Many animals have "played possum" to convince another party that they were dead. The phrase pays tribute to the playacting skills of the opossum, an unobtrusive little marsupial with a reputation for being, mentally, rather a dim bulb. But the opossum does put on a superb show when threatened. When danger approaches, the opossum drops in a tight ball as if feigning death, a good move since many predators will leave a carcass alone. It may also defecate, producing an odor that heightens the illusion of a dead animal. An opossum can hold this position for hours, especially if nudged with a stick. When not in danger of being eaten itself, opossums will make a meal of almost anything, including eggs, baby birds, small rodents, frogs, fish, insects, full-grown chickens, and snakes. In fact, they suffer no ill effects if bitten by a rattlesnake, cotton mouth, or copperhead, probably because their snake-eating habit has helped to build a strong evolutionary pressure to develop resistance to the venom.

The only marsupial found north of Mexico, opossums are about the size of a honeybee at birth. Equipped with strong forelimbs, they drag themselves along their mother's fur and into her pouch, where they fasten onto a mammary gland. Eight weeks later they emerge to spend their next month clinging precariously to the mother's fur as she moves around. After that, they are on their own. Opossums are recognizable by their white faces, dark gray fur, black eye patches, and pointed snouts. Recently, they have been spreading in the United States, largely as a result of human transport. Without doubt, the opossum has toughed it out superbly in a world dominated by placental mammals.

For the Love of Honey

The African honey guide loves honeycomb and isn't shy about trying to get it. On locating a beehive, the little bird will find a mammal—a honey badger, baboon, or even a human—and try to attract its attention with special cries and aerial maneuvers. Often, the helper will follow the bird a kilometer or more to the prize. If the bird is lucky, the helper will open the hive and take most of the honey but leave enough for the honey guide to feast on. The honey guide harbors bacteria in its gut that digest the beeswax and release the calories for the bird's use. Honey guides thus need helpers—mammalian and microbial—for starting and finishing each beeswax banquet.

Animal Defenses

Striking with blinding speed, a snake seizes the tail of a small lizard. But instead of a whole lizard, it ends up with just the rear part of the tail, which breaks off as soon as the snake tugs. The snake has had the misfortune to attack a green anole, a lizard whose tail is built to break and leave predators with only a souvenir of a meal that might have been. Elsewhere, a house cat takes on a porcupine and soon finds itself covered with sharp, barbed quills working their way into its flesh. On the way home, the cat comes face-to-face with a dog. Unable to run away, it arches its back, bares its teeth, and stands its hair on end to make itself look as big as possible. The dog gives the cat a wide berth.

These encounters illustrate just a few of the innumerable strategems that animals use to confuse, frighten, or out-hustle a would-be predator. Some animals flee, some camouflage themselves, and still others use poison, sharp spines, or other nasty surprises to ward off unwelcome pursuers. Even big, tough animals with no fear of predation must sometimes defend their young. Witness a herd of musk oxen set upon by wolves intent on devouring their calves. The oxen will form a circle, with the calves inside, and it is the wolves' task to try to break it open. As prey species evolve better defenses, predators also develop better means to overcome them, and so is nature's game of cat and mouse constantly refined.

Some animals aren't at all subtle with their weaponry. Certain tropical fish, some ostentatiously colored and others nearly invisible against the ocean bottom, pack their poison in their dorsal fin spines. The gorgeously striped lionfish and the gaudy red, blue, and green scorpionfish store potent venom there. Even more dangerous is the stonefish, a dark, bottom-dwelling reef fish that looks like a rough rock. A diver who steps on a stonefish may face days of agony or even death if the poison-laden dorsal spines are driven into a heel or instep. Many small invertebrates employ elegant chemical defenses. Among recent discoveries: Milkweed bugs taste so bad that a praying mantis can easily learn to avoid the bugs or even a harmless beetle painted to look like one. And millipedes secrete a slow-acting Quaaludelike chemical that sedates wolf spiders on the attack.

Scorpions make deadly use of the poisonous stinger at the end of their tails, but the whip scorpion uses a slightly different approach. The whip scorpion resembles the scorpion, but its tail ends in a long, thin, nonvenomous tip. Its owner can aim the whip in any direction, releasing a noxious spray that irritates the eyes or other susceptible tissues of a predator.

Some animal populations permanently change their coloration

to blend in better with their background. When the Industrial Revolution blackened trees in England with soot, light-colored moths were more readily caught by birds because they stood out against the dark trunks. As a result, more dark-colored moths survived and the population shifted toward darker colors in polluted areas. In nonpolluted areas, however, light moths remained hard to see, and the population didn't undergo the shift. Proper coloration may also mean survival to guppies; studies indicate that female guppies have evolved a preference for drab males over brightly colored ones in waters where the guppies are preyed upon by large numbers of predator fish.

Many animals hide when danger approaches: The turtle or tortoise withdraws into its shell, the prairie dog heads for its burrow, and the mouse may stand stock-still so as not to betray itself by movement. The spiny anteater, or echidna, of Australia is covered by a coat of spines similar to a porcupine's. The echidna also has powerful claws and can dig itself a hole in the hard earth in just a few minutes. Crawling in, it leaves only a bed of spines on the surface for the intruder to see. A tiny architect of stream beds, the caddis-fly larva builds its own shelter from grains of sand (or other hard bits of material) that it glues together, to form a stony cylinder enclosing its long, slender body.

Camouflage provides another means of hiding. From brown "stick" insects that resemble twigs to green katydids taken for young leaves, animals have evolved to appear inconspicuous. A type of caterpillar spins dried feces into a camouflage coat. The decorator crabs, a subgroup of spider crabs, actively disguise themselves by covering their shells with so much algae, sponges, bryozoans (tiny "moss animals"), or other living organisms that a predator or person can miss them entirely. Some go an extra step by including in the decorations distasteful sponges or stinging hydroids, which can repel predators.

The hognose snake, a harmless variety, uses a bit of mimicry in resembling its dangerous kin, the rattler. But if that fails to deter a predator, it can bloat itself to nearly twice its normal girth and hiss. Failing that, it may play dead as convincingly as any human actor, complete with limp body, lying on its back, and tongue hanging out. This last posture, if it is not to be the snake's final action, must at least blunt the hunting instinct of a predator conditioned to strike only at moving, and therefore living, prey.

All of these mechanisms evolved over eons of give-and-take struggle between predators and prey or between animals and the elements. In the future, naturalists are sure to unearth even more startling examples of defense, particularly among smaller, less-studied creatures, such as insects, that for millions of years have waged a battle for survival on a miniature scale. Those war stories are just beginning.

Nature's Behemoth

In the heyday of whaling, no whale fetched a higher price than the sperm. Rich in oil and slower than the bigger blue whale, the "cachalot"—named from the Gascogne word "cachau," or "big tooth"—was hunted relentlessly by whalers who threatened its existence. Sperm whales were targeted largely because of the fifteen or so barrels of top-quality oil that they carry behind their great blunt foreheads. That oil apparently gave the whale its name, since whalers once mistook the oil for sperm. The only great whale with teeth, the sperm was chased in every sea and was hunted until 1989. It is easy to identify not only by its shape but by its blow, a three- to five-meter cloud of water vapor projecting upward and forward from the top of its head.

Whalers once told harrowing tales of rowboats overturned and main vessels attacked by wounded sperms. These exploits survive in whaling scenes carved on sperm-whale teeth, an art form called scrimshaw. Despite the brutal image of sperm whales that they portray, no records exist of unprovoked sperm-whale attacks on humans. On the contrary, it was whalers who showed the sperm no mercy, even when it displayed traits of loyalty and defiance that would do humans credit. On one occasion, a harpooned sperm was encircled by other sperms, who faced away from their wounded companion and slapped their tail flukes on the water as if making a stand in its defense. Apparently unmoved, the whalers took advantage of the whales' immobility and proceeded to harpoon every last one.

The role of the oil in the sperm whale's head isn't known, but scientists have speculated that it may be needed for buoyancy or as part of an echolocation system for locating prey. If the latter is the case, the oil may form an echo chamber or other sound-related structure. Evidence for echolocation is good, but not yet airtight. The whales emit a series of clicking sounds, about one per second, as they cruise, and reflections of these signals may help them home in on dinner. Researchers listening in on whales have heard the clicks speed up, presumably as the whales were closing in on their victims, followed by the cessation of clicking, a sign that the prey had been seized. Since the whales do a good deal of their hunting at night, sonar would make an ideal tool.

However they find it, sperm whales need a daily diet of about 800 pounds of squid and fish to maintain their 12-ton bodies.

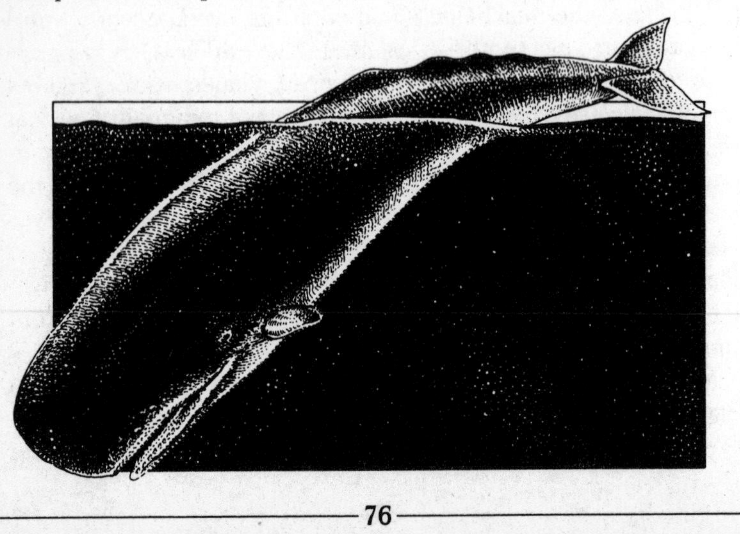

Squid seem to be the whale's favorite food. Souvenirs of encounters with the 10-armed invertebrates crop up in sperm-whale feces, which almost always contain lots of horny squid beaks (some from giant squid up to ten meters long), and on the whales' heads, which often bear scars from squid suckers. In searching out their prey, the whales dive to incredible depths; one was found drowned and entangled in a cable set two-thirds of a mile deep off South America, and circumstantial evidence suggests they can dive to 10,000 feet.

Wear and tear from diving and fights with giant squid notwithstanding, sperm whales may live up to 70 years. The largest individual on record was a male just over 60 feet; the largest female was about 55 feet. Breeding takes place in relatively warm water, and gestation lasts 16 months. Calves subsist on rich mother's milk for about two years, after which they catch their own food. At about five years of age the males wander away from their mother's pod to form their own pods, while the females stay with their home group. As they continue to grow, the young males move into progressively colder waters and end up spending most of their time in cold polar or subpolar oceans, returning to warm seas once a year to mate. Adult females, on the other hand, spend their entire lives within about 40 degrees of the equator.

The blocky-looking body belies the sperm's wonderful grace and aquatic skill. Underwater it maneuvers with ease, and at the surface it will play with floating logs, seaweed, or other objects. Full-grown whales in Antarctic waters have sustained dives for up to an hour; off Japan the record is 30 to 40 minutes. In calm water the buoyant whale may sleep at the surface, either lying prone or hanging upright but in any case with its blowhole in the air. Usually the sleeper is accompanied by a few other whales keeping watch.

While the head produces high-grade oil, the other end of the sperm whale holds its own in the valuable product department. A thick, waxy substance called ambergris (for its amber-gray color) is found near the end of the whale's large intestine. Like feces, it contains squid beaks and other food residues, but its origin isn't known. It may be formed as a result of constipation, or perhaps because of one of the peculiarities of the mating season, when adults, especially sparring males, cut down on eating and form a food residue with a low feces content. Used as a stabilizer in perfumes, ambergris varies in quality and price but should be prized by any beachcomber lucky enough to find some washed ashore.

However valuable, ambergris and the large quantities of oil and meat obtained by slaughter cannot compete with the long-term benefits of keeping sperm whales alive. Having them around to watch and study will surely prove more worthwhile than the products of their death.

Why Do Some Beetles Have Horns?

If you can't figure out why a beetle would sport a set of horns to shame a rhinoceros, you're in good company. Charles Darwin wondered the same thing about some Chilean beetles that he found during his voyage on the *Beagle.* Of course, the beetles' horns are smaller, but in proportion to their size they would outmatch any moose's or elk's. Until fairly recently, scientists could not be sure what the beetles' horns were for, since they seemed incapable of hurting a human finger or even other beetles. Now close observations have revealed that many species do use their horns in fights with others of their kind, but in ways no one had guessed.

Unlike the antlers of deer or the horns of bulls, the weapons of beetle gladiators seem designed to pry an opponent up and off a twig or leaf. Or, when the horns work like pincers, the beetle may pick up its adversary and drop it to the ground. In most cases the horns inflict no injury but simply get the other beetle out of the way so that the victor can enjoy access to a mate or a piece of real estate.

Beetle horns come in a bewildering array of shapes and sizes. Some grow up from the head, others from the mandibles, various parts of the thorax (the beetle's second body region), the front legs, or the wing covers overlying most of the abdomen. In the case of one darkling beetle, two horns project upward from the mandibles and end in pointy knobs, reminiscent of an upside-down golf club. When two beetles meet head-on, the horns fit nowhere. Since the beetles live in narrow tunnels in logs where maneuvering space is scarce, it was thought that the horns could not play a role in combat. But this idea was turned on its head—literally—when scientists found that in an encounter the more aggressive beetle tended to take to the ceiling of a tunnel, still facing its opponent but now with its horns pointing down. In this position the horns can indeed pinch the opponent, and the knobs can pierce a membrane between the head and thorax, drawing blood.

In another species of beetle, the male burrows into sugar-cane stalks and attracts females with a pheromone. If a rival challenges the resident male, the two fight for possession of the burrow. Each is equipped with two horns, one sticking back from the head and the other curving forward from the prothorax

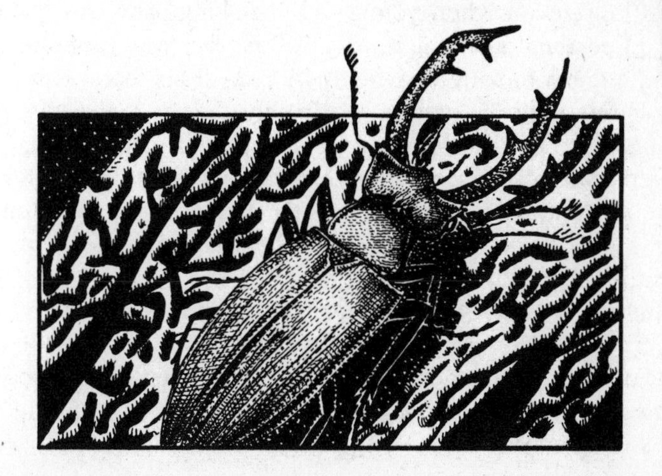

(the front part of the thorax). The resident, with his rear to the burrow opening, must turn and face the challenger to vanquish him. In turning, he must hang upside down from the burrow's upper lip, facing inward. As he hangs, his rival butts him repeatedly from below with his rear horn. Finally, he is allowed to complete the maneuver and face the challenger. Each then tries to put his head horn below and his prothoracic horn above the other's head, grasping it in a pincerlike grip. The winner will be the beetle that manages to lift and flip the other off the stalk.

How Shocking Are Electric Eels?

The South American electric eel can deliver quite a wallop, even when young. An inch-long baby can make a person's skin tingle; a yard-long eel can generate 350 volts, enough to stun a horse; and a five- or six-footer can "fry" its victims with 650 volts. Presumably they can also dispatch people quite easily, as shown by a couple of incidents that occurred many years ago. On both occasions, men drilling for oil near the upper Orinoco River were found drowned in small ponds. The men were good swimmers, and it was presumed that they died from electric eel shock.

Any animal with a nervous system produces electricity, but the tiny currents are confined to the internal organs. Electric eels, on the other hand, send shock waves outside their bodies. Armed with three electricity-generating organs—two producing weak currents and one strong—they swim along, creating weak electric fields around them. If a fish comes close enough to distort the field, the eel turns on its main organ to stun or kill it. The electric organs enable the eel to "see" objects in the muddy water around it, gradually replacing the eyes, which go blind as the eel ages, perhaps from being exposed to so many electric shocks.

The electric organs occupy the tail section, which accounts for four-fifths of an eel's body. The other vital organs are all crammed into the front, making it a short fish with a very long tail. Actually, electric eels are not true eels; they just look eel-like thanks to their long bodies and finless backs. Another oddity: The electric eel must breathe air every fifteen minutes or so or it will drown. Lacking lungs and possessing nonfunctional gills, it absorbs oxygen through numerous fleshy folds inside its mouth. Thanks to them an eel can be kept out of water for hours, as long as its mouth is moistened regularly.

Miniature Sharks

The spined pygmy shark, reaching only about 9.8 inches, wins the title for the smallest shark, with the pygmy ribbon-tail cat shark running a close second. People rarely see the spined pygmy, because it passes its days 200 meters or more deep on the sea floor, rising only at night to feed near the surface. Its belly contains luminescent organs that help conceal it from predators below. A note on the pygmy ribbon-tail: It gives birth to young about four inches long, even though the mother may be only six inches long herself.

Poison-Packing Amphibians

Clearly visible in the surrounding foliage, the dart poison frog's bright pattern, whether red, orange, or other bold colors, sends a clear warning to predators. The little frogs carry a battery of skin glands that secrete toxins so deadly that South American Indians "milk" the glands for poison with which to tip their arrows and darts. One bite—from the predator, not the frog—can result in death. Some frogs, from Colombia west of the Andes, make a poison stronger than curare. The plodding cane, or marine, toads use a similar defense. Native to South America but introduced to such places as Florida, Hawaii, and Australia, these toads can reportedly weigh up to six pounds and eat frogs, lizards, mice, and maybe even small birds. A dog that bites a cane toad can end up temporarily paralyzed or dead from poison secreted by glands on the toad's back. Reports of people licking the backs of cane toads have recently surfaced. Taken in small doses, the poison is said to produce psychedelic hallucinations.

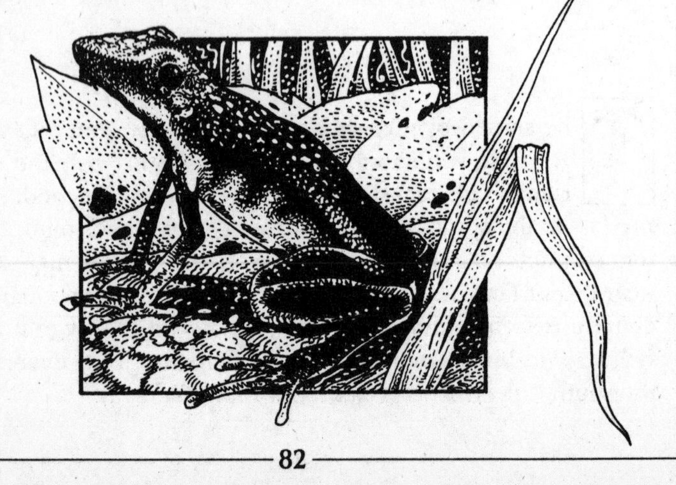

Snail Warfare in the Pacific

To a gourmet, one of the great pleasures in life is *escargots,* or snails, a mainstay of French cuisine. So it is understandable that people saw fit to introduce the giant African land snail, whose meaty bulk supports a shell up to twelve inches long, as a food source for Pacific islands from Hawaii to New Guinea. Unfortunately, bigger has not proven better for the islands where the huge snail settled. The African giant ate crops, invaded homes, and generally made a nuisance of itself. In an effort to undo the error, carnivorous snails from Florida and Central America were shipped to the islands, where, it was hoped, they would make short work of the giant snail. The carnivorous snails went right to work and gobbled up every other snail in sight—except the African. They devastated native snail populations everywhere, leaving hardly a survivor to be found. A particularly sad case was on Moorea, one of the French Society Islands, near Tahiti, where about seven native snail species had been studied for a century and provided the basis for much work on evolutionary mechanisms. The Florida snail wiped them out, but not before scientists had rescued some individuals and whisked them off to zoos in the United States, Europe, and Australia, where they are being bred for eventual release in their homeland—if the carnivorous snail dies out. So far, that snail shows no sign of relinquishing its hold on the island. None of it would have happened had people not shortsightedly introduced the African snail without studying the consequences. Thus, while it may be unfair to blame the African directly for the native snail deaths, it goes to show that even a vegetarian species can wreak havoc with other species by prompting humans to act unwisely.

Walking Fish

Sometimes even fish feel the need to go for a stroll. Ponds dry up or become low in oxygen, rich food sources beckon on land, predators approach—the reasons are many. But aside from the problem of breathing air, land travel means forsaking the near-weightlessness of being in water for an unsupportive environment where fragile fins are nearly useless. While many fish have found a way to negotiate land, the only ones that have become really skilled at fin walking are the mudskippers. Also known as walking gobies, mudskippers live near mud flats and mangrove swamps in parts of the Indian and West Pacific oceans and tropical West Africa. They come ashore at low tide to feast on worms, mollusks, insects, and small crabs, spotting their prey with highly placed eyes that can look in dif-

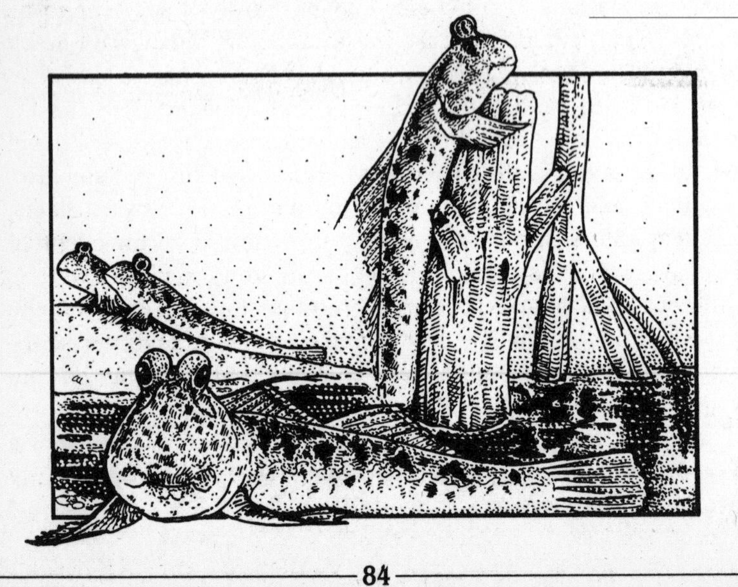

ferent directions at once. The mudskipper walks with the aid of pectoral fins anchored to strong bones by a muscular lobe. The lowest fin rays in the pectoral and tail fins are unusually tough, as are all the pelvic fin rays. To walk, the fish holds its weight on the pelvic fins and draws the lower pectoral rays together to make a solid support, then plies the pectorals like oars. In this manner it clambers not only over mud, but rocks and protruding mangrove roots. Mudskippers also "hop" across the mud by bending their tails and sinking the firm tail-fin rays into the mud. Then, straightening the tail with a powerful snap, they propel themselves forward, perhaps helping with a boost from the pelvic fins. Other fish with a similar hop include the sheepshead molly miller of tropical eastern North America, certain tropical blennies, and walking catfish. But for sheer fin power, the mudskipper walks away with the honors.

The Greatest Masterpiece of Animal Architecture

Truly a wonder of the world, a great wall, built over thousands of years by countless tiny animals, stands out to an astronaut's eye. Stretching 1,200 miles off the northeast coast of Australia, the Great Barrier Reef dwarfs all others and deserves to be called the greatest work ever produced by nonhuman architects.

Australia did not always have the abundance of coral it has today. Millions of years ago it was attached to Antarctica and experienced a much colder climate. But the Earth's crustal movements broke it away from Antarctica, and it drifted north, into warmer seas, where coral could grow. The modern reef arose from the latest round of growth, which began about 8,000 years ago.

The reef lies approximately 40 miles out from the mainland. Its largest coral mass is about 45 miles wide, and only about ten routes through the reef are safe to navigate. It comprises a series of smaller reefs and coral islands that rose toward the surface as coral skeletons accumulated. Corals are simple little animals related to jellyfish and sea anemones, but unlike their cousins they secrete a hard, calcium-rich cement that anchors them to one spot and forms a matrix in which many individuals live as a colony. Succeeding generations of coral larvae settle on the old framework and add their own matrix, piling up the cement until the reef almost reaches the surface. Individual coral animals are called polyps, a term referring to a stage in the life cycle of coral and their relatives. (Jellyfish represent another stage, called a medusa, which coral never goes through.) The polyps feed much like anemones and jellyfish, using tentacles set with stinging cells to paralyze prey.

Corals of the Great Barrier Reef will grow only between temperatures of 65 and 96 degrees Fahrenheit and thrive between 77 and 86 degrees. They cannot live in murky water. Parts of the reef reach to a depth greater than 1,000 feet, but the living polyps are rare below 150 feet. The reason coral prefer clear, shallow water goes beyond the fact that it's warmer near the surface. Algae are commonly found living inside the cells of the polyps, and they need light for photosynthesis. The arrangement works out nicely, for the polyps get energy produced from the algae's photosynthesis, and the algae get nutrients from wastes produced by the coral. In fact, without algae there would be no reef. Coral needs the energy supplied by algae to produce enough calcium carbonate cement for a reef to accumulate. Thus the mighty Great Barrier Reef owes its existence to a partnership of two "lowly" life forms: tiny coral and microscopic algae.

Coral colonies grow into a vast array of shapes. Brain coral grows round, with folds that resemble the convolutions of the human brain. Staghorns grow up in stately columns; branch coral bifurcates like the branches of trees; flange coral spreads out flat like pancakes; and others form delicate networks like a mass of twigs. The colors vary, too, depending on the pigments in each species of polyp. In all, more than 340 types of coral, or more than half the species found in the Pacific and Indian oceans, grow in the reef.

Coral form three main types of reef, all of which occur in the Great Barrier Reef complex: barrier reefs, or breakwaters encircling a lagoon and islands; atolls, or round reefs enclosing a lagoon with no island; and fringing reefs, which stem from coral growth on a flat sea bottom near an island or mainland. Sometimes outcroppings of coral will form a cay, or small island, when they trap sediment and floating vegetation, which lays the basis for soil.

Besides coral, the reef harbors countless other life forms, including about a thousand kinds of fish. Some have evolved astounding living strategies, some are deadly, some are beautifully simple, and some are simply beautiful. Their presence makes the reef the natural wonder it is, rather than just a string of coral heaps.

The two-foot unicorn fish certainly deserves its name. A prominent horn protrudes from above its eyes and points toward its mouth. The parrotfish, with its beaklike mouth, looks a good deal like its feathered namesake. Gorgeously patterned fish like the orange-striped harlequin tusk fish and the blue-, black-, and white-striped emperor angelfish share space with equally dazzling sea slugs, crabs, mollusks, worms, and starfish.

The underwater garden is sprinkled with colorful, flowerlike sea anemones, their tubelike bodies topped by a crown of stinging tentacles. No flower but a carnivorous animal, the sea anemone paralyzes small fish, crustaceans, or other animals that become entangled in the tentacles and thrusts them down through its mouth to be digested. Yet the porcelain crab makes its home within and suffers no harm. Male and female clown fish nest in their host sea anemone without fear for their offspring. The male tends the brood right under the canopy of tentacles, throwing out any infected or infertile eggs he may find.

The male cardinalfish also broods eggs, but in an even more protected spot: his mouth. He makes up for the relative lack of oxygen in this unusual brood chamber by periodically spitting the eggs out about halfway to replenish the supply. The male triggerfish builds a saucer-shaped nest several yards wide from chunks of coral. The male keeps it free of sand so the eggs, when laid, will have clean surfaces to attach to. He defends the nest against rival males by engaging them in mouth-to-mouth combat, which may leave both fish with torn lips. When the moon reaches the first quarter, female triggerfish appear and the male tries to lure one down to inspect his nest. Courtship begins at twilight, followed by egg-laying at dawn. The female tends the nest until the eggs hatch the next night. The young of the black and white damselfish receive even more parental attention. Whereas most fish leave the hatchlings to fend for themselves, the female damselfish shepherds her babies like a mother duck. Some damselfish have another peculiar habit, related to food rather than reproduction: They cultivate their own private patches of algae and defend them from other fish.

In November, the southern seas warm up, and the reef coral prepares to mate. A few days after the full moon the waters turn cloudy as polyps shoot millions of tiny packets of eggs and sperm out through their mouths. The thick flurry blankets the entire reef as the corals, cued by the moon and water temperature, all release their packets at once. The bundles float to the surface, where they break open and liberate the eggs and sperm, and fertilization occurs.

While the corals are busy renewing themselves, other species go on making life difficult for them. One coral cruncher, the hump-headed parrotfish, bites off big chunks of coral and ingests the polyps, spitting out the inedible matrix as sand. Another, the crown-of-thorns starfish, moves slowly over the reef, leaving denuded coral skeletons in its wake. The crown of thorns can occasionally get out of hand and wreak havoc with coral, as it did in 1962. It has been thought that these outbreaks resulted from human disturbance, but recent evidence suggests that the crown of thorns has inhabited the reef for at least 8,000 years and may even have contributed to its diversity of species. Even so, these starfish pose a menace to unwary divers, since they grow as broad as 18 inches and sport up to 17 venom-spined arms.

The reef's denizens display a bewildering variety of colors or camouflage, depending on whether their strategy is to warn off predators or escape their notice. The brilliantly striped lionfish, with its poisonous spines, serves warning to meal-minded bigger fish, but the dull-colored stonefish, equipped with spines to pierce and kill anything that steps on it, spends its time blending in with the bottom rocks. A gaudily colored nudibranch mollusk, shell-less and carrying its gills on its back, will poison a predator. The ribbon-shaped young batfish wriggles through the water looking like a bad-tasting flatworm but will mature into a striking fish, with prominent dorsal and ventral fins.

Some harmless creatures look fearsome. Neither the huge whale shark nor the spotted leopard shark can bite a bather. The giant manta ray, black-backed and white-bellied, has earned the nickname "devilfish" but can make a good friend to human divers. Even the dreaded moray eel has been befriended by divers visiting its stony lair. In contrast, the innocuous-looking cone shells can fool a beachcomber accustomed to harmless snails and hermit crabs hiding in shells. Cones kill their prey with a poisonous dart-tipped tongue and can severely wound or even kill a human. Colorful giant clams, their shells a foot or more across, sit with open valves, filtering out bits of food from the sea water, waiting for prey to wander by.

On the cays, giant green turtles come ashore in December to lay their eggs. The turtles may be a yard across and 500 pounds, but their hatchlings make easy prey for herons. The mother turtles themselves make things difficult for others of their species, digging their nests without consideration for other nests and sometimes scattering other turtles' hatchlings with their flippers.

By night the beaches are covered with ghost crabs, skittish crustaceans that run for their burrows at the first hint of danger. They can see in all directions, thanks to vertical eyestalks topped by cylindrical pupils. The crabs scavenge bits of food left behind by other animals during the day, but they have also been known to kill newly hatched turtles.

The Great Barrier Reef is home to other life forms too numerous to mention. It is one of the world's most precious treasures, but it has been abused by humans. Runoff from farms on the mainland has emptied fertilizer and other pollutants into the pristine sea, and guano mining that began in the late nineteenth century nearly devastated an island at the southern tip of the reef. The establishment in 1979 of Great Barrier Reef Marine Park in the southern reef is a welcome sign of protection for the coral and the life that it supports. If left alone, it will continue to evolve and be shaped by the forces of waves, land movements, and biological activity that have made it one of the wonders of the world. Its corals stand as a living reminder that small and lowly creatures can be the mightiest of all.

The Waddling Parrot

A huge parrot that can't fly—this image barely begins to describe the kakapo of New Zealand, surely the world's oddest parrot. Abundant when the Maori people settled New Zealand and when Captain Cook visited in 1770, the kakapo has been decimated by animals that arrived with people, especially cats, rats, and mustelids (relatives of the weasel, introduced to control rabbits). Predominantly green but with a sunburst of brown bristles on either cheek, which give it an owlish appearance (it was once called the owl parrot), the kakapo has short wings but very strong claws. It climbs trees to feed on berries, buds, and other vegetation and uses its wings to glide down to the ground. According to some reports, however, when frightened, it may simply drop off a branch without unfolding its wings and hit the ground like a dead weight. It long ago lost the ability to fly, a result of living in the isolated, predator-free New Zealand islands. On the ground, it walks with a waddling, ducklike gait along paths through the vegetation. Measuring two feet from beak to tip of tail and weighing up to five and a half pounds, this largest of parrots is nocturnal and shy. Mating appears to take place at special places where the males have hollowed out little depressions in the ground. Standing in them, they sing out their deep, booming mating calls, which carry a long way. The female, who alone tends the chicks, lays two or three eggs in a nest built in the shelter of aboveground tree roots—not a very secure place if a weasel is around. Since she must leave the nest to feed at night, this multiplies the danger to the young. When cornered, an adult kakapo can put up a terrific fight against foes as big as dogs, even to the point of wounding them seriously with its beak and claws. The New Zealand Wildlife Service has made great efforts to save the kakapo by moving the few specimens left to islands free of predators, but suitable places are rare, and stoats (a type of mustelid) have at least once swum to an island refuge and attacked the parrots. The plight of the kakapo typifies the predicament of species confined to islands: If predators are introduced, habitat is lost, or disease strikes, they have nowhere to go. For the earthbound kakapo, it's like living in a birdcage in a house full of cats.

A Real-Life Dragon

A visit to the remote, nearly inaccessible Indonesian island of Komodo would convince the most die-hard antiromantic that dragons belong to the real world as well as to fairy tales. There, and on the west part of Flores Island, live about 5,000 specimens of the Komodo dragon, the world's largest lizard, a dark-skinned hulk with a very reptilian temper. The Komodo dragon possesses a keen sense of smell but poor eyesight and hearing. Growing as large as ten feet long and more than 200 pounds, it eats young goats and buffalo and is quite capable of killing a person who gets too close. It can also outrun the fastest humans. Attracted by the smell of carrion, which they can detect from as far as seven kilometers with a favorable wind, several dragons may converge on a single carcass, in which case the biggest individual will get most of it. That gives them all a powerful stimulus to grow fast, as does the fact that adults sometimes cannibalize juveniles. If they survive to adulthood, Komodo dragons may live as long as 80 years.

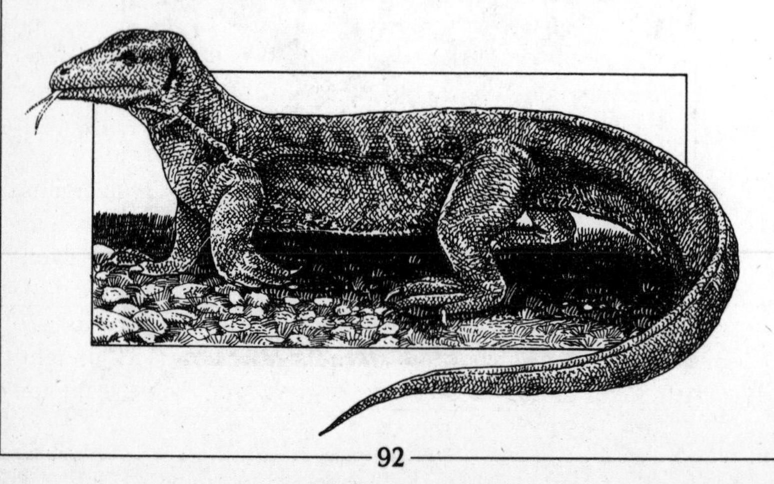

Partnerships between Species

Anywhere that organisms come into contact there's a chance that they will develop a living arrangement in which at least one depends on the other for some or all of its needs. Common examples include plants that depend on insects to fertilize their flowers and the intestinal bacteria that enable cows to digest cellulose. In some cases, both organisms benefit; in others, one benefits and the other suffers no harm; while in still others, one organism derives an advantage at the other's expense. In many instances it is difficult for scientists to sort out which hurts or benefits which; it is easier, therefore, to refer to all such arrangements as forms of symbiosis, from the Greek for "living together."

Lichens, the plants that first colonize rock and prepare the way for its breakdown by mosses and higher plants, consist of an algal and a fungal species living in close contact. The fungus can't live on the rock by itself, but it derives nutrition from photosynthesis performed by the alga. The alga seems to benefit from the fungus's ability to retain water, yet it can live on its own. Plants known as legumes, which belong to the pea family, contain root nodules housing bacteria that convert gaseous nitrogen into forms that the plant can use. All are examples of symbiotic living.

Several deep-sea fish harbor bacteria in special light organs that may attract or confuse prey. The bacteria produce light continuously by chemical reactions, and some fish can even blink by means of a dark skin flap that works like an eyelid. A giant clam common to coral reefs keeps brown algae in its gill tissues and consequently must leave its valves open to sunlight for their sake, though it probably derives food from its photosynthetic companions. Reef fish, plagued by parasites, regularly visit "cleaning stations," where small "cleaner" fish or shrimp glide among their scales, into their gills, and into their mouths—even the mouths of fierce sharks—to feed on the parasites or remove dead tissue. One scientist found how important the cleaners were by removing all cleaner fish from a reef area. Soon most of the customer fish were gone, and the remaining few appeared weakened. Although cleaning is mutually beneficial, cleaners don't always depend on customers for their entire diet, and some cleaners may be eaten by their customers.

Some fish that thrive where others dare not go include the little damselfish, which lives among the tentacles of sea anemones, and a brightly colored fish associated with a floating relative of the sea anemone, namely, the Portuguese man-of-war. Both appear to lure larger fish to their death among the stinging tentacles of their "hosts" and then feed off "crumbs" from the feast. On the ocean bottom, a segmented worm lives in a U-shaped burrow that it shares with "boarders," including crabs or scale worms. The host worm pulls a current of water through its burrow and feeds off particles that it filters out; the boarders evidently feed on leftover tidbits.

The acacia tree of Central America houses an unusual ally: ants. The ants make their homes in the thorns of the tree and, in return, defend it against all comers, attacking anything that touches it, including other insects and people. The leafcutter ants of tropical American forests diligently tend underground gardens of fungi, feeding off the fungal crops when they mature. Other ant farmers "milk" aphids for a sweet liquid called honeydew.

The tuatara lizard, found on a few islands off New Zealand, shares a nest with the sooty shearwater, an often harmonious arrangement. The bird forages by day and the lizard by night, so they don't get into each other's way much. When the shearwater leaves to migrate, the tuatara has the burrow to itself to hibernate. A much larger reptile, the Nile crocodile, allows plovers to enter its open mouth, where the birds fearlessly feed on leeches attached to its gums. Also in Africa, little tick birds hitch rides on the backs of rhinoceroses and feed on ticks clinging to the rhinos' thick skin. The tick birds also warn the rhinos of danger with a chirp—a great help to the huge mammals, which have notoriously bad eyesight.

Humans have their share of symbionts, too. One of the most obnoxious is the Chinese liver fluke, a parasitic flatworm that causes serious problems in parts of the Far East that lack modern sanitation. Another, the protozoan that causes malaria, lives and reproduces in red blood cells. We also have some friendly symbionts; the most notable are the intestinal microorganisms that make vitamins B_{12}, E, niacin, pantothenic acid, and folic acid.

Perhaps the greatest cases of symbiosis occurred eons ago. Evidence exists that the cells of higher animals and plants evolved with the help of microbes that invaded the cells and turned into organelles such as mitochondria and chloroplasts. Both are membrane-bound structures that occur outside a cell's nucleus; the mitochondria are the powerhouses that produce much of a cell's energy, and chloroplasts perform the vital function of photosynthesis. Without those invasions we wouldn't be here, or we would at least differ from our familiar selves. What other feats of symbiosis are still to come?

Skin-Breathing Vertebrates

It may seem odd to contemplate, but a large number of vertebrates already equipped with perfectly good lungs or gills do much of their breathing through their skin. Fish that spend time on land may exchange up to half their total respiratory gases this way. In fact, skin-breathing may have been an important crutch that helped ancient vertebrates colonize the land. Modern amphibians are quite good at it. The Lake Titicaca frog, native to that large body of water between Peru and Bolivia, is draped with huge folds of skin around its trunk and hind legs. The folding so increases the surface area of the skin that the frog doesn't need its lungs to absorb oxygen and release carbon dioxide. The male of another species, the hairy frog, grows fingerlike projections of skin over its rear trunk and hind legs during the mating season. The increased surface area helps the frog keep up with the heavier respiratory load imposed by the stress of looking for a mate. Perhaps for the same reason, the anal fins of male South American lungfish sprout feathery projections of skin, richly supplied with blood vessels, during its mating season. Fresh-water and marine snakes do varying amounts of skin-breathing. Some sea snakes may even "blow off" nitrogen through their skin during long, deep dives as a hedge against getting the bends when they surface. Other animals that rely heavily on skin-breathing include the mudskipper, a fish of mangrove swamps that spends much time out of the water; the hellbender, a wrinkly skinned, two-foot-long salamander of the eastern and central United States; the flatfish known as the plaice; bullfrogs; goldfish; several turtles; and the lungless salamander, which breathes exclusively through its skin. In addition, some bats take advantage of the thin membranes in their wings; the big brown bat eliminates up to 12 percent of its carbon dioxide this way. Also, the skin of young birds is often richly supplied with blood vessels and likely plays some role in respiration, more so than adult skin. Even humans skin-breathe a little bit. Our skin cells exchange respiratory gases with the air, accounting for a few percent of our total exchange. It's a good thing, however, that skin-breathing hasn't replaced lung-breathing in our species, for without lungs we would have neither speech nor song.

Legless Lizards

The tropics are home to the most unusual lizards on earth: the worm lizards, a group that, except for three species, has lost its legs. Soil burrowers all, they subsist on real worms and insects. The Florida worm lizard is often mistaken for an earthworm, and for good reason; it looks almost exactly like one, except for its size. Its foot-long, sinuous body moves through moist soil like an earthworm; its pale skin is divided into short segments by a series of rings along its entire length; and it has no functional eyes or ear openings. It burrows with help from its shovel-shaped snout. The ajolote of Baja California and two related Mexican worm lizards haven't gotten around to dropping all their legs, but still could cause a casual observer to do a double take. The ajolote has tiny front legs, complete with claws that help in burrowing; tiny, dark eyes; and, like the Florida worm lizard, skin segmented by deep ringlike lines. The ajolote's skull and face, however, are much more lizardlike. The

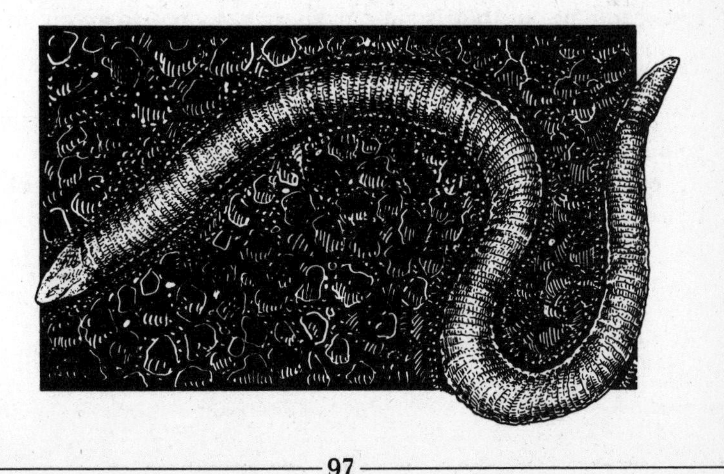

face lacks rings but is scored by a pattern of deep lines reminiscent of cracked mud on a dry lake bed. Surely the ajolote and its cousins win few points for attractiveness, but if one is going to live like a worm, it helps to be built like one, too.

Deafening Frogs

For sheer cacophony, the Puerto Rican coqui frog is practically in a class by itself. The male of this two-inch-long brown species produces a ko-*kee!* sound as loud as 108 decibels, enough to compete with a jet engine passing low overhead. The males sing out their name in hopes of attracting a female, usually at night but in some areas after every rain shower. The "ko" part of the song seems to indicate a warning to other males to stay away, while the "kee" invites females to join the singer. The males, equipped with powerful trunk muscles, can belt out their song quickly in response to an outbreak of quiet in the jungle. Within a quarter-second of noise cessation, the frogs can start their calls. The call is actually somewhat pleasant to human ears, although a chorus of thousands at once could make for a sleepless night in the Puerto Rican rain forest. In recent years, the coqui's numbers have declined noticeably, however—a setback suffered by many other frog species around the world.

Lizards That Curl into "Wreaths"

Covered with brown spiked scales, the South African armadillo lizard looks like a miniature crocodile at first glance. A crocodile's scales lie flat, however, unlike the rough and spiky lizard's. Usually a sluggish mover, the lizard often can blend in with sand, thanks to its dull color. But let danger threaten and the armadillo lizard heads for a crevice, where it clamps its tail in its mouth and curls up in a ball. This position protects its vulnerable underside and turns the armadillo—also known as the wreath lizard—into a good imitation of a dried wreath. The tail that serves the lizard in such good stead remains in place for the next emergency. Unlike some other lizard tails, the wreath lizard's is firmly attached and doesn't readily break off when a predator—or its owner—pulls on it.

The Biggest Animal—Ever

Dwarfing all but the largest dinosaurs, the blue whale stands—or floats—unrivaled as the most massive animal that has ever lived. Although the largest dinosaur found to date was longer, the blue whale outweighs it by a large margin. The dinosaur, a 47-ton behemoth tentatively called seismosaurus, measured about 140 feet from snout to tip of tail. In contrast, the largest blue whale known was 98 feet, and weighed about 190 tons, and many individuals have grown to the neighborhood of 100 tons.

In the water, the blue whale cuts a striking figure. Even its tail end is magnificent, set off by flukes nearly 20 feet across, about twice as wide as a sperm whale's. Baby blues start life 20 to 25 feet long, as big as killer whales. Nourished by rich mother's milk, they pack on 200 pounds a day; at weaning they switch to a diet consisting almost exclusively of small crustaceans called krill. The species' great size poses no hindrance to speed; in fact, they were rarely hunted until the late nineteenth century, when steam power and exploding harpoons finally gave whalers the advantage. Catches peaked at close to 30,000 in 1931, and blues have been protected from whaling since 1966. Still, this

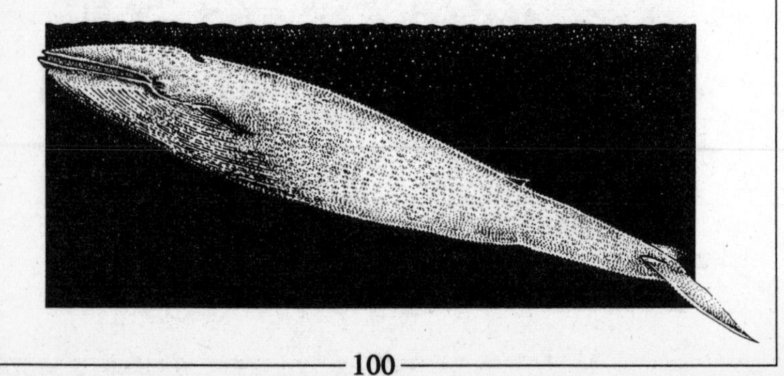

majestic creature could be slipping toward extinction. An International Whaling Commission survey of blues from 1978 to 1979 and 1985 to 1986 in their summer feeding grounds near Antarctica led to an estimate of only 500 individuals in an area where commission scientists had expected to find almost ten times as many. Such a dismal count indicates that the blue whale may be unable to rebound from hunting as fast as once thought. Unfortunately, its size and four-ton-a-day krill appetite make survival all the harder. Perhaps with continued protection the whale will be spared, at least for the near future, the fate of the dinosaurs. If not, the planet will lose the biggest treasure it has ever possessed.

How Smart Are Dolphins?

Over the years dolphins have acquired a reputation as some of the most intelligent animals on earth, ranking with apes. The lovable, beak-nosed little whales possess large brains and learn tricks with remarkable ease, but assigning them a secure place in the intellectual realm has proved difficult. Dolphins live in a world different from ours, and any animal's intelligence is geared to its own particular niche. Still, the 30-plus species of dolphins do seem capable of many feats beyond their wild repertoire and should be credited with a certain amount of cerebral wattage. For example, dolphins are among the most playful of creatures and have been quick to pick up the rules of games played with people. Trained dolphins can understand simple sign-language commands and score well when the "words" appear in an unfamiliar order, signaling a new instruction. One trainer even got her dolphins to master a simple abstract concept; when she rewarded them with fish only for new tricks, they seemed to grasp the concept of "newness" and proceeded to invent a succession of movements. There is no doubt that these animals are well-endowed mentally, but some animal trainers claim similar behavior for dogs and possibly even pigeons. As we get to know our sleek cetacean friends better—which will take some effort—we will gain a better understanding and appreciation of their intellectual gifts. Perhaps they will, too, of ours.

Avian High Divers

Among all the waterfowl and sea birds that aerial-dive for their dinner, it may be unfair to pick a champion. Brown pelicans plummet to the water and scoop up fish in their expandable, shovellike jaws and huge pouches. Terns skydive with uncanny accuracy, some staying dry as they snatch fish in their bills and others immersing themselves in the effort. Kingfishers perform elegant headfirsts into the water, as do some gulls, shearwaters, and skuas. And don't forget the penguin, which leaves its terrestrial ungainliness behind as it gracefully swan-dives from an ice ledge into the cold Antarctic seas.

But consider the feats of boobies, goose-sized tropical and subtropical sea birds, and their relatives the gannets. Spectacular high divers all, they differ in technique and in the types of prey that they pursue. The red-footed booby dines largely on squid and flying fish and has been seen deftly pursuing flying fish in the air. The masked booby dives from heights estimated at up

to 100 meters and penetrates the water very deeply. Brown boobies, and probably others, chase fish underwater by swimming with their feet and wings. The male blue-footed booby, with his very long tail and small size, can dive from great heights into less than half a meter of water. Blue-footeds also have developed a communal hunting strategy, in which an entire group dives at once when one of them signals by whistling. Gannets, large and strong enough to handle formidable prey such as big mackerel, also plunge from on high, sometimes corkscrewing and sometimes hurtling straight down. Both boobies and gannets get help from air sacs under the skin, which cushion the impact and help buoy them up. Such artistry as theirs surely earns these remarkable birds a "booby prize" for high dives.

One Hundred Years in the Desert

Somewhere in the Mojave Desert of California, two male desert tortoises meet by chance. They wag their heads up and down and stare for a while, sizing each other up. Their weapons are simple: thick shells up to 15 inches long, strong forelimbs, and, jutting forward beneath their chins, a projection of the plastron (the under-armor) called the gular horn. They charge and begin to fight, each trying to hook his gular horn under the other's shell and tip him over. Finally one succeeds and plods away, leaving his vanquished opponent to right himself as best he can.

The loser pushes at the ground with his forelimbs and neck. If he is small enough, he will probably manage to flip back upright and may go on to live out the four- or five-score-year life span that his species is heir to. If he fails, he will soon heat up and die in the desert sun. His death would be a loss to California, which needs all the tortoises it can keep. So do Utah, Nevada, New Mexico, and Arizona, where the reptiles also range. South of the border, in the Mexican state of Chihuahua, the California's cousin the *bolsón* tortoise faces a similar challenge. Like its northern relative, the *bolsón* is threatened by human development and is literally trying to stay on its feet in the struggle to survive.

Excellent diggers, both tortoises excavate burrows in which they wait out the winter or hide from the blazing summer sun. Humidity in the burrows exceeds that of the surrounding desert air, giving the tortoises a hedge against desiccation. Also, the two tortoises seem able to absorb water from their bladders. But none of this guarantees safety.

In the western Mojave, the desert tortoise is up against a foe that thrives in the developed areas that the tortoise shuns. Ravens have followed people into the desert, drawn by garbage and sustained by their ability to nest on fence posts and power-line towers. Beneath the nests, piles of baby tortoise shells bear mute witness to the raven's appetite for the young reptiles. Just a few decades ago the adult *bolsón* suffered a similar fate at the hands of human hunters. Its ability to detect the ground vibrations of approaching people didn't protect it much. Ducking into its burrow at the first sign of danger, it would often find itself pulled out by a pole stuck down the hole and hooked under its shell. At up to 18 inches and 30 pounds, it is the largest of the North American tortoises, an ample supplier of meat for tortoise stew.

Today, tortoise hunting is illegal and people are making amends. Mexico has set up a tortoise reserve and hatcheries where females are brought and given hormones to induce egg-laying. As many as 80 percent of the tiny hatchlings survive, to be released into the desert, a far cry from the 2 to 3 percent that make it in the wild. California, too, has established a reserve, and federal officials have made efforts to control ravens. An adoption program allows about 20,000 pet tortoises to live the life of Riley with their owners, who give them more care than a dog or cat, including putting them to "bed" for their annual four-month dormant period. The flip side of such domestic bliss is that the tortoises apparently learn to recognize their adoptive family and become helpless without them, a potentially fatal situation for an animal that's released or wanders

back into the desert. A former pet that's used to being fed, for example, may not forage for itself. Or, accustomed to the family dog, it may neglect to retreat into its shell at the approach of a predator. From an owner's viewpoint, a pet tortoise means a serious commitment. Since the tortoises may live 80 to 100 years, it can be a long one, too.

Desert and *bolsón* tortoises are thriving in reserves and homes, but they aren't out of the woods yet. Civilization continues to encroach on the borders of their habitat, while off-road vehicles threaten them in the interior. An epidemic respiratory infection has been found in the desert tortoise, and it is still too early to predict either species' long-term fate. Only continued vigilance will enable future generations to live out their long lives in their desert abode.

Nature's Sanitation Brigade

Since time immemorial the plains of Africa have supported vast herds of elephants and grazing animals. They, in turn, have blessed the grasslands and savannahs with countless tons of droppings every day. Yet the plains are not drowning in dung; on the contrary, hardly any is to be found except what was left in the last day or so. How does it disappear so fast?

Most of it falls to the "shovel patrol" of more than 2,000 species of dung beetles. These insects, some tiny and some titanic, pick up the smell of fresh dung and make a beeline upwind toward the pile. A contingent of beetles can dispose of an eight-gallon pile of elephant dung in a day, eating some and feeding the rest to their offspring. Without them the entire continent would soon choke in its own waste.

Most dung beetles work at night, in relative safety from predators, but a few are diurnal. The most famous are the scarab beetles that rim the Mediterranean and were sacred to the ancient Egyptians. The Egyptians likened the scarab, continually rolling a ball of dung that it had formed, to the force that rolled the sun across the sky. Other beetles break up the dung by burrowing into it or by carrying bits of dung down tunnels that they have excavated in the soil below or next to the pile. In the tunnels, the beetles mold the dung into balls into which the females lay their eggs.

The ball-rollers appear to have the most arduous task. These beetles pick a bit of dung sticking out from the pile, burrow around it to cut it off, then pat it into a neat ball, roll the ball as far as 15 meters from the pile, and bury it. The ball-rollers must also look out for thieves that will try to steal a ball. Fights over balls happen fairly frequently on dung piles, but injuries almost never occur. In some species both sexes make balls, and in others the female joins a male who is already making one. The female may help the male by pulling on the ball as he pushes it, or she may follow behind him. She may even ride the ball, looking a bit like a lumberjack birling on a log, as it turns over beneath her. The speed of rolling also varies with the species; the champions can roll their dung balls up to 14 meters per minute over level ground. After the ball is interred at the burial site, the pair mate and leave eggs to hatch into larvae that will feed on the dung.

However it processes dung, speed is of the essence to the dung beetle. Hungry birds, mongooses, and other animals are attracted to beetle-infested dung piles, and a slow beetle may find slim pickin's if many others have beaten it to the pile. A single mass of elephant dung may contain upwards of 7,000 beetles, busily sorting out the dung components. The beetles discard inedible bits of roughage and eggs of parasitic worms and flies, leaving a thin mat of coarse fibers, where only hours before lay a deep pile of dung. The parasites' eggs perish or leave only a few deformed offspring. Thanks to the beetles' efforts, few eggs of insect pests such as bush flies survive. Their burrowing also loosens the soil, improving it for any seeds that fall from the dung and increasing its water-holding capacity.

Dung beetles also live in the Western Hemisphere. The first beetles successfully transplanted for the control of dung and associated pests were taken from Mexico to Hawaii in 1923. Afro-Asian beetles were subsequently introduced there, and in 1967 African beetles were introduced into Australia, whose native beetles prefer marsupial to cattle dung. The immigrant beetles have done well at controlling such bovine pests as horn flies. In Australia, especially, they have alleviated problems with dung, which not only accumulates but fosters the growth of plants unpalatable to cattle, unfortunately decreasing the amount of pasture available.

Is the Grizzly Bear as Fearsome as Its Reputation?

In mid-January 1970, a wilderness guide named Harvey Cardinal made a startling discovery. In the woods near Fort St. John in northern British Columbia, a big grizzly bear was still wandering in the snow, not yet hibernating. Eager to claim its valuable pelt, head, and claws, Cardinal set out to hunt the bear. Following its tracks, he passed a mossy hummock about five or six feet high, never suspecting that the bear was lying behind it. As soon as he passed, the bear circled behind the hummock and came up behind him. Cardinal died about six feet beyond the hummock, his gloves and his rifle's safety still on. The tracks in the snow told the story of a cunning ambush and a victim who had no warning.

That kind of tragedy doesn't happen often; only 41 people are known to have died from grizzly attacks between 1900 and 1980. But when it does happen, death by grizzly is about as terrifying as it gets. Grizzly bears are unpredictable and quite capable of acting out the stereotype of savage killers. On the other hand, the bears generally avoid humans. Several attacks can be put down to a sow defending her cubs or a bear so hungry that it overcame its natural wariness of humans. The great brown bear is less tolerant of people than its black cousin, and when bear and people—that is, society's interests—clash, the bear always loses.

The history of grizzly-human interactions has not been happy. As white ranchers settled the West, the grizzly and its taste for livestock put it on the ranchers' hit list. And hit it they did, with bullets, traps, and poison. State by state, the grizzly began disappearing after the turn of the century and now exists mostly in Yellowstone and Glacier national parks and a few other pockets of the lower forty-eight. Of the perhaps 40,000 to 60,000 bears in North America, only several hundred live in the the continental United States; the rest inhabit Alaska and Canada.

The grizzly belongs to a species that includes the Alaskan brown bear and the Kodiak bear. Its gray or silver-tipped fur gives it a grizzled look that, along with a hump above its shoulders, distinguishes it from lighter-colored black bears. It arrived in the New World by the same route that people did—a land bridge across the Bering Strait that once connected Alaska and Siberia. The bears crossed over about 50,000 years ago and gradually migrated as far south as Mexico.

A good-sized adult bear may reach close to 1,000 pounds, but 200 to 600 is more common. Whatever the size at maturity, all grizzlies start out as helpless cubs weighing only about a pound. Cubs are born in midwinter, while the mother is busy sleeping her winter sleep. They stay with her for about two years, until they can fend for themselves. A sow usually mates every third year, beginning at age five or six, and continues into her twenties. Mating takes place in summer, after the two- or two-and-a-half-year-old cubs strike off on their own. As in black bears, implantation into the uterus is delayed until the female seeks out a den in the fall.

Unlike black bears, though, the grizzly sow brooks no interference with her cubs. While a few people have carried off screaming black-bear cubs without being attacked by the mother, woe to any person or beast that threatens a grizzly cub. One notable exception, the boar grizzly, occasionally makes a cannibalistic meal of cubs and will attack sows to get them, sometimes killing an entire family. Often, though, the sow's ferocity makes up for her smaller size and convinces the boar to retreat and seek food elsewhere.

Despite the grizzly's prowess as a fighter, it doesn't make a career of hunting game. Rather, it gorges on nuts, berries, roots, and other vegetation, taking game when the opportunity arises. Its victims include squirrels, rabbits, deer, elk, and moose; the calves of hooved mammals are a favorite item. The appeal of grizzly cubs to predatory boars may stem from their relative heft, as walking meals go, and the fact that once the mother is out of the way, bear cubs are usually much easier to catch than agile rodents.

Since summers don't last very long in most grizzly habitats, the bears must stuff themselves with enormous amounts of food to put on enough fat to last the winter. Thirty thousand calories a day, or about ten times an adult human's energy intake, is not unusual. During the six months of denning, adult bears slowly metabolize the fat and abstain from eating and drinking. The tiny cubs, however, need to grow fast in order to emerge from the den strong and ready to travel with their mother. They nurse her rich milk and play with their siblings—one, two, or three of them—to pass the time.

In early spring, the bears leave the den and being feeding again. Many of their first meals consist of carcasses of animals that died in winter. Later, they turn to game. One of the most fascinating sights in Alaska is the salmon spawning runs that draw bears to shallow rivers, where they fish with their paws. For fishing bears, salmon are a gold mine of protein, fat, and calories needed for fattening up. A hungry bear can easily devour 15 big fish in a day. Spawning runs often provide such a cornucopia of food that the bears, normally jealous of their food supply, don't seem to mind sharing space with one another or with curious humans. That isn't to say that a bear will understand if a person reels in a fish that the bear fancies. Fishermen plying the streams are advised to cut their lines and let the bears have the fish rather than risk a dispute. It makes little sense to pick a fight with an animal capable of sprinting up to 35 miles per hour and killing a person with one swipe of its paw.

The power of grizzlies almost defies belief. After killing a thousand-pound elk or buffalo, a bear may pick up its prey in its jaws and move it to a less conspicuous location where other bears are unlikely to intrude on its meal. Big game cannot be consumed at one sitting, so bears often hide their kills and return to them later. They have been known to bury a carcass partially and cover it with sticks and leaves to prevent either sight or smell from attracting a potential rival.

This little bit of foresight hints at another grizzly trait: their high intelligence. From the affection of a sow for her cubs to a young bear's playful batting of a snowball, grizzly behavior has much in common with that of dolphins, dogs, and other intelligent animals. A grizzly adopted as a cub can grow up much like any other oversized pet, namely, fond of people and trainable, though with a mind of its own. And, of course, its appetite would make a Saint Bernard look like a piker.

Tame or captive animals have little to fear from people; it's the wild ones that are endangered by their very wildness. The bear that has all but disappeared from its southern range could become threatened in the North, too, if development pressures continue. Human settlement has cut off the Yellowstone and other national park populations from one another, a situation that leaves very little margin for misfortune. Wild animals cannot long exist in small tracts of wilderness, and even the largest national parks may not be big enough. Aside from the plague of inbreeding and subsequent weakening that small populations are heir to, small habitats can be decimated virtually overnight by fire, drought, disease, or any number of other catastrophes. If the population cannot shift to new lands because it is hemmed in, then it will suffer the consequences to the full.

In facing the problem of how or whether to protect the grizzly, people must inevitably choose the kind of world they want. The grizzly represents nature in her most pristine form: terrifying, powerful, and untamed—the embodiment of the universal human nightmare. Yet how many of us really want a world where no such terror exists? Do we also want a world devoid of wild lions, tigers, and wolves? The grizzly has always been a part of wilderness, which is to say, a part of our heritage. Soon society will have to decide whether to allow true wilderness, including the grizzly, a place on our planet. If not, we will book future generations on a voyage of no return toward a world where all nature is as safe, benign, and controlled as a golf course.

The Fiercest-Looking Pig

Fierce pigs aren't hard to find. People of Africa and the Eurasian continent, for example, fear such swine as wart hogs and giant forest hogs, especially the latter, the male of which is apt to charge without warning. But the fiercest-*looking* pig is probably the babirusa of Indonesia, a strange, rather tame species that is often domesticated. Its face, however, looks almost demonic at first glance. The light-colored eyes glimmer menacingly against the pig's dark, brownish-gray skin, and the four tusks, which develop to the greatest extent in males, stick up like scimitars. The bottom tusks, like those of most wild pigs, grow up from the sides of the jaw, but the upper tusks grow through the skin at the top of the muzzle—never entering the mouth—and curve back toward the forehead. The tusks seem to help the males in fighting, the uppers generally for defense and the lowers for offense. So large and prominent are they that natives liken them to the antlers of a deer. In fact, babirusa

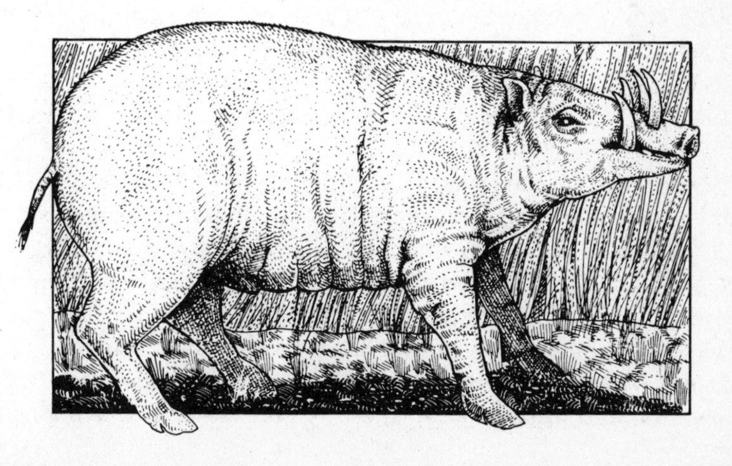

means deer-pig or pig-deer. As one Indonesian legend has it, this pig hangs itself by its tusks from a tree branch every night. A swift runner weighing up to 200 pounds or more, the babirusa prefers moist habitats and sometimes swims across short stretches of ocean to get to new islands.

Spider-Eating Insects

Creepy, crawly, and menacing, spiders would make any Who's Who of unpopular creatures. Pity the poor insect tangled up in a spider's web, helplessly waiting to be pierced and sucked dry. In real life, though, insects sometimes turn the tables. Even the biggest spiders are no match for certain wasps, who feed the living spiders to their young.

Such a wasp is the tarantula hawk, a name given to several North American species that take on the granddaddy of spiders. The wasp seeks out tarantulas because she must sting and paralyze one to provide a source of fresh meat for the larva that develops from her egg. The wasp-spider duel is a fight to the death, as fascinating as it is furious, especially for anyone who has shuddered at the thought of grappling with an oversized tarantula.

Depending on the species, the tarantula hawk may be an inch or an inch and a half long—very large for a wasp, but still about eight times smaller than an average-sized adult tarantula. Larger than the male, the female alone engages tarantulas in the dance of death. To build strength for the great task before her, she feeds voraciously on nectar or fruit juices. When ready to lay an egg, she flies around in search of tarantula burrows. The spiders spend much time in these steeply sloping excavations, which they dig several inches into the soil. The entrance may be webbed over, a sure sign that the occupant is at home. When she finds a burrow, the wasp does everything she can to entice the spider out. She may tug on strands of web or pat the soil, but if the tarantula doesn't oblige, she will go in after it. The majority of encounters, however, appear to take place on open ground.

The battle begins when the spider rushes from its hideout, perhaps expecting to find a tasty fly, only to come face-to-face with its mortal enemy. Sometimes the tarantula runs away and escapes the wasp. If not, it rears up and tries to push the wasp away. She presses the attack, trying to swivel her abdomen around to sting the spider in a vulnerable spot. The wasp has an advantage in keener eyesight and a tough cuticle that the tarantula's fangs cannot always pierce. The tarantula can strike with lightning speed, but it doesn't see very well up close, and the size and number of its legs—eight compared to the wasp's six—don't help much.

The struggle ends, usually, when the wasp's stinger finds its mark in the spider's underside. The spider crumples, limp and completely at the wasp's mercy. Seizing it in her mandibles, she drags her prize to the hole that she has selected, which may be a burrow she has dug or perhaps an old rodent hole. Pulling the living but paralyzed spider into the burrow, she lays a single egg on its abdomen. Leaving the burrow, she covers the opening and goes about her business. She may fight several more tarantulas before the summer is over.

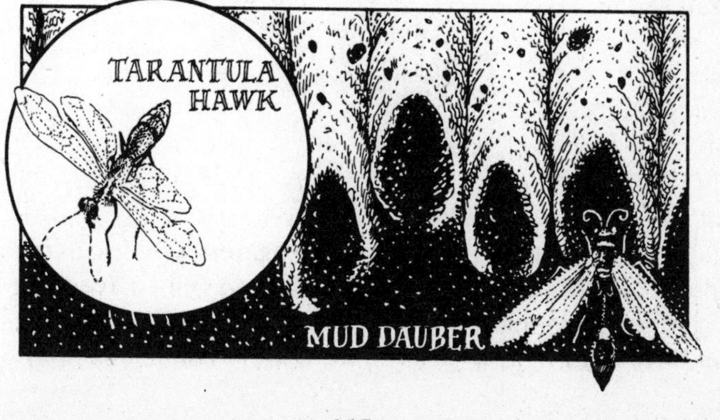

Meanwhile, the developing larva consumes first the tarantula's abdomen, then its cephalothorax, and proceeds to spin its cocoon over the spider's remains. The whole process takes about a month. If the egg fails to develop, the tarantula may live in its tomb for two months, until it dies of desiccation.

Another wasp that parasitizes spiders is the dirt dauber (or mud dauber), a species common in eastern North America. Its name comes from the tubular brood cells that the females construct out of dried mud. These gray nests resemble pipe organs stuck vertically to the sides of houses, barns, garages, bridge tunnels, or similar structures. Each "pipe" is a cell that will be stocked with spiders to feed a single egg. A nest usually contains two to five cells stacked one on another.

After she has completed her nest and the brood cells are sealed off except for one end, the female flies off in search of orb-spinning spiders. After stinging a spider, she bears the paralyzed prey in her mandibles back to a brood cell, packs it in, and goes looking for another spider. When several have been packed in the cell, she lays an egg on the living pile of provisions and seals the entry. She repeats the process until all of her brood cells have been stocked. The larva consumes its food in about a week, then spins its cocoon. On emerging from the cocoon, the young adult chews its way out of the brood cell.

Killer Bees

In 1956, 26 queen bees of a very aggressive African variety, along with attendant drones, escaped from a breeding laboratory in Brazil. Since then, these bees have spread rapidly north to the southern border of the United States. Extremely

defensive about their hives, they will sting anything that disturbs them and have already killed thousands of animals and several people. The African bees carry no more venom than the familiar European honeybees, owing their deadliness instead to the great swarms that attack at once. Known popularly as killer bees, these African bees nest in the ground or take over managed hives and make better pollinators than the European bees, but produce much less honey. Biologists had hoped that as the bees spread, their undesirable traits would be diluted by interbreeding with European bees, but the African variety has displayed an astonishing ability to breed only with members of its own kind. Hence, the bees heading north are almost pure African. Since they are tropical, some biologists speculated that they would have difficulty surviving the cold winters in the northern United States; they have proved quite capable of surviving in the mountains of Argentina and Colombia, however, and an African queen has even made it through a German winter. So far, no one has found a way to stop their progress north. It looks as though we're destined to be plagued with them, defenseless except for a very elemental tactic: to run at the first sign of danger. A queen won't kill you, but her "subjects" will die trying.

The Forbidding Mangrove

If ever a tree species has been unjustly maligned, it is the mangrove, the centerpiece of mangrove swamps. Mangroves once ringed about 60 percent of all tropical coastlines, but destruction by humans has lowered the figure to less than 40 percent. Certainly these remarkable trees deserve better. They have conquered a harsh, salty environment to provide a haven for birds, a nursery for some of the most economically valuable oceanic creatures, and hunting grounds for such mammals as monkeys, raccoons, and even tigers.

From a boat, the edge of a mangrove swamp looks like a bowlegged chorus line marching out to sea. The trees' eerie appearance comes from the "prop roots" that grow down from living mangrove trunks and branches, splaying out like legs as they near the ground. The prop roots penetrate seawater and anchor themselves in mud, bracing the trees so they can remain upright despite their unstable watery base.

The key to the mangrove's importance lies in the rain of leaves it drops into the shallow water. Once released by bacteria and fungi, nutrients from the leaves fuel a tremendously productive food chain that nourishes the young of many species, including pink shrimp, tarpon, and snappers. Without this shoreline cradle, shrimperies and fisheries all over the world would suffer, some severely. Its tangle of roots and mud offers ideal protection for nesting herons, pelicans, ibises, egrets, roseate spoonbills, and other wading birds. At low tide the predators that roam the mud flats looking for crabs or other invertebrates find a rich feast.

Besides the trees themselves, some very striking species make their home in mangrove swamps. In the swamps bordering India and Bangladesh, Bengal tigers hunt for small mammals and may take human victims unlucky enough to wander into their territory. The long-tailed macaque, a crab-eating monkey, searches the mud flats of Southeast Asia for its favorite crustaceans. A second primate, the proboscis monkey, lives only in the mangrove swamps of Borneo, where it dines on mangrove leaves.

The value of mangrove swamps extends beyond the species it shelters. Mangrove forests help protect coastal areas from storm battering and produce materials such as wood for making paper, tannins for curing hides, and charcoal. If managed and harvested prudently, they can coexist peacefully with humans. It only remains for governments and individuals to recognize their importance and the fact that these forbidding swamps are vital to some of the species we most cherish. In this case, a lack of beauty is no gauge of worth.

A Rock-Climbing Fish

Pity the poor salmon that can't make it up a high waterfall without a fish ladder. If only they could take a page from the rock-climbing catfish of Colombia, they would make the spawning run a lot easier on themselves. The catfish isn't much to look at and isn't a very graceful swimmer, but it can scale the face of a waterfall without a rope. Its underside has all the equipment that it needs. First, its sucker mouth latches onto any surface and stays stuck, even against a torrential flood. Second, a bony plate on its belly enables it to hold its place as it loosens and repositions the sucker. Attached to the plate are ventral fins fitted with tiny, sharp, backward-pointing "teeth" that dig into the rock and prevent slipping. With powerful muscles controlling the plate, the fish is doubly secure as it lifts and advances its sucker. By alternately moving sucker and plate, the catfish can "walk" its way up a 20-foot perpendicular waterfall in about 30 minutes.

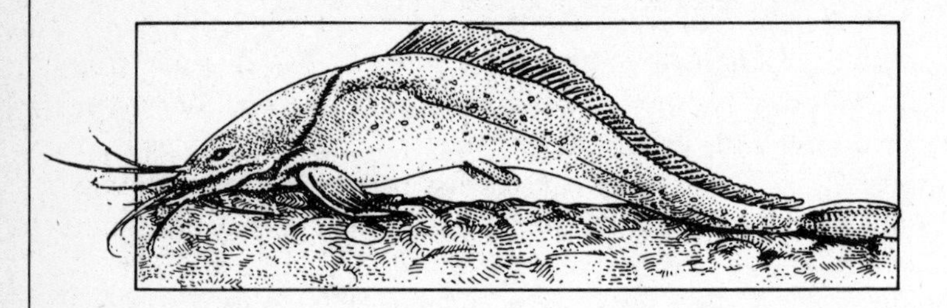

An Amphibian Ballet

A graceful duet is the last thing one expects from the Pipa toad, a flat-bodied amphibian with a head that appears to have been stepped on. But this toad, which never leaves the water, performs a well-choreographed mating ritual that begins with the male sitting on the female's back and grasping her under the arms. The female then kicks, sending the pair up into a lazy somersault. On the way down, she releases a few eggs and the male releases some sperm to fertilize them. He then uses his webbed hind feet to shepherd the eggs onto the female's back, where they stick. Over and over they repeat this aquatic pas de deux, until about 100 eggs have been spread on the female's back. The eggs soon disappear beneath the surface of the skin, emerging 24 days later as tadpoles.

Horrifying Fish

Sharing the globe with fish of striking beauty, ferocity, or oddity are some that can only be termed disgusting. The faceless hagfish, for example, can tie itself in knots and secrete enough slime to slip from the tightest grasp. The lamprey bores its circular, rasping mouth into the sides of bony fish and eats their living flesh. But for sheer skin-crawling loathsomeness, consider some members of a tiny South American catfish family. Only about an inch long and toothpick-thin, these fish are among South America's smallest. But what they lack in size they make up for in nuisance. Some species burrow into the gills of bigger fish, where they suck blood from the rich network of vessels that pervade the gill slits. Even more bloodcurdling are several species of the candiru, which—not often, thank goodness—attack both men and women as they bathe in streams. These fish wriggle their way up the urethra, attach themselves, and suck blood. Backward-pointing, erectile spines along their heads act like the barbs on a harpoon and make it impossible to extract the fish by pulling. They can cause agonizing pain, and the only way to get rid of them is by surgery.

Underwater Castles

The vast family of cichlids, which come in many varieties, exhibits some of the most remarkably diverse behavior in the world of fish. For example, several African species are sand-castle builders and may cover the shallow bottom of a big lake with an architecture of different buildings, some seven feet high, others ten feet across, each with a sand platform for mating. During the mating season these males become the biological version of sand shovels, continuously building castles. These castles are similar to the bowers of bowerbirds, functioning as a lure for females. Cichlids are known for their feisty territorial behavior. Indeed, mating season turns a lake, at least the two-and-a-half-mile-long sand-castle area of 50,000 bright-blue males (observed in one lake), into a war zone.

Another of the cichlid species dances before mating, whereupon the female lays her eggs, gathers them into her mouth, then nuzzles the male's anal area to receive the sperm into her mouth for fertilization. Some of these species can carry up to 20 fry this way until the young are ready to leave home, watch-

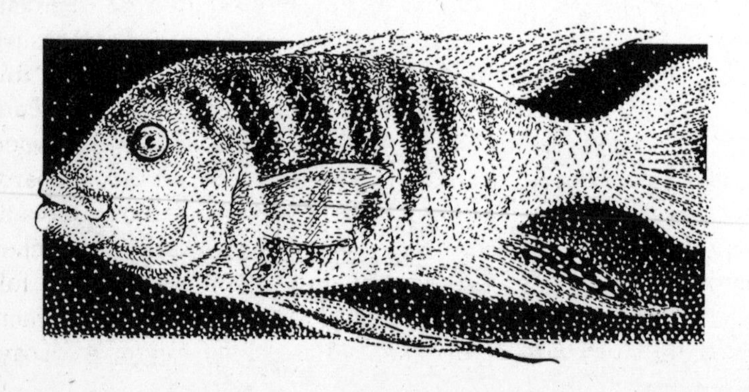

ing out all the while for yet another species—the head rammer—which eats babies after ramming the mother's head.

A third variety of cichlid, a brown and white one, will suddenly fall into the weeds and turn white. Since it looks like a corpse, other fish will come to eat it—but just as one approaches, up it jumps to eat them.

Still another species takes over empty snail shells, each male holding a harem of females in an array of the old shells. And if he should die, one of his wives will literally become a male to take his place.

And there are others, one called "fat lips" because it sucks food out of cracks in the rock, another with a large, curved beak for eating around corners, and those who specialize in nibbling the fins and scales and even eyes of other fish. Cichlids are truly a zoo unto themselves.

The Most Dangerous Flower

Don't fall onto this flower: Rigid spikes decorate its center, it smells like rotting flesh, its bloom can measure easily one yard across, and it weighs 15 pounds in full bloom. The bunga patma (genus *Rafflesia*) is, in fact, truly beautiful. Growing in parts of the rain forest of Sumatra and Borneo and on two islands of Indonesia, it displays its dramatic leopard-like inner colors and curled petals for only four to six days at full blazing bloom. First visible as a new bud only two inches across, the flower requires about nine months to grow to full size. At that point it entices its first visitors: The smell attracts flies, the inner parts squirrels. Soon the bloom begins to decay,

dissolving to black slime within a few weeks. After another seven months, a mushy fruit filled with seeds appears. This attracts deer, wild pigs, and more squirrels. All these visitors may well help in its pollination—and the plant is *not* carnivorous at any stage—though this theory, and many others about the flower, have not been verified.

The bunga patma is known to be a parasite, though. Lacking roots or the ability to draw nutrients from the sun, it grows from tiny seeds that fall into wild grapevines tangled on the forest floor. With no visible leaves or stems, it seems to emerge directly from the bark on which it grows. The plant draws all its nutrients from the vine's stems, yet does not seem to damage the grapevine.

Bunga patma, a local name, means "lotus flower," and it is regarded as a symbol of fertility. The bud extract is prescribed in folk medicine to women after childbirth. For this reason, and especially because the rain forest itself is in danger of destruction, this magnificent flower's survival is endangered. May it always bloom, strange but grandiose.

Luminescent Insects

From the flicker of fireflies over warm, barefoot grass on a June evening to the firefly trees of Asia, lighting up dark rivers at night as millions of the tiny creatures flash in synchrony, firefly light is perhaps the most exquisite sexual invitation in the world of insects. And it is rather common: There are more than 2,000 species of these small, soft-winged, nocturnal beetles, many of which make light. The bright glow-worm is a larval stage.

The firefly's lovely green-gold flicker is conjured in its abdomen as a substance called luciferin is oxidized by the enzyme luciferase. The light is merely a by-product of this chemical transaction, a brief release of energy. The "fire" is actually cool, containing virtually none of the infrared wavelengths possible within the spectrum of light, and nearly all visible, containing almost no ultraviolet components. Attractive to other fireflies as well as to us, it makes them one of the few insects anywhere to use sight instead of smell to find mates.

The "roving fireflies" of the northeastern and midwestern United States are solitary, each searching independently for a mate. From dusk until the fall of total darkness or later, the males fly flashing until their signal is returned by a female lying in low grass or bush. Their light is about twice as bright as hers. The two dance a duet of light until mating. By morning they are dark and hiding in the grass.

From India through New Guinea other, more social, species of fireflies are found, those that flash synchronously in vast groups within tall, full trees. The natural Christmas trees they create, often several in a row along a river, can actually light a moonless night well enough to read by. Their coordination, based on precise feedback within their nervous systems, has evolved in places where a bright, roving courtship would be almost impossible because of the thick tropical vegetation. This elaborate group behavior could probably never have evolved unless it first somehow served the individual firefly's success in producing offspring—this is the major route by which evolution works. And single males do indeed show their own drive by flashing more intensely when landing in a new place and encountering other males; they can also direct their light signal individually by twisting their bodies. But the bright synchrony has also come to help every firefly in the group see a mate more easily than it would otherwise. And it does not reward the individual insect at the expense of the group, since the female firefly's eyes are sensitive only at certain intervals. This means that no single male can gain an advantage by flashing off-rhythm to draw attention to himself. So ragtime is not rewarded in the social fireflies' world, but instead a simpler, unsyncopated song. Among the fireflies of North America, the flashing may have been refined by death: The female flies do eat up some males who fail to flash in a way attractive to their mating sense.

This kind of living light, a type of bioluminescence, is not confined to fireflies. Many other insects glow—either from luminous bacteria in their bodies, a luminous secretion, or with a metabolic by-product similar to the firefly's. One click beetle of the tropics even produces both red and green light. Fungi have it, too, with some toadstools glowing a faint green at night in the woods, for example. And bioluminescent creatures are even more numerous in the deep ocean than on land. The sea is decorated with flashlight fish and related species of many kinds.

A phenomenon this pervasive probably has more than one evolutionary impetus. One theory is especially curious: It may be that using up oxygen was the impetus for the firefly's synthesis involving it. Before there were plants to make oxygen, our planet had very little of it, and early creatures had not evolved to need it. They found it, in fact, somewhat poisonous. The firefly's use of it to synthesize light, an advantage, remained because it didn't hurt and soon helped, the theory goes.

Although no one would insist upon a human benefit from the firefly's existence, there is one. Their luciferase is an enzyme that can help in screening for human tumors and urine infections, in testing for blood problems, and as a fast-acting detector of an infection or of ingredients in various medicines; it has been on the market, in a genetically engineered form, for a year or so already. But for every person who benefits in this way, there are probably thousands who simply love the light.

The Subtle Serpent

Somewhere in the world, about once every 50 years, a human baby is swallowed whole by a large snake. It might be a python, which can gulp down a small impala of almost its own weight, or an anaconda, another constrictor that grows up to 37 feet long, or even a 4-foot rattlesnake, since they can swallow full-grown cottontail rabbits.

The snakes do it with their lower jaws and other related adaptations. The jaw can separate into left and right sections joined by only a stretching ligament, with both parts only loosely attached to the skull. Since the ends of the ribs are not attached to a breastbone, this allows flexibility for even more stretching room. The sides of the snake's body and the walls of its throat and stomach are made of soft, elastic skin, which also aids in the stretch. A bone case for its brain prevents that organ from being dangerously shoved, and a snake can even breathe quite easily when its mouth is stuffed full of prey because its trachea leads straight to the nostrils and has no connection to the mouth. No wonder the Bible chooses the snake for its starring role in Eden and dubs it "more subtile than any beast of the field."

And this is not all. The snake family is also amazing in its ability to squeeze a prey to death—boa constrictors and anacondas are particularly good at this—and in its eerie talent for sensing the heat given off by living prey, as well as in its method of injecting venom.

The heat-sensing ability of two families of snakes—the vipers and the constrictors—enables them to find their prey even in total darkness. The heat detection organ, one or more pits on either side of the head, jaw, snout, or under the eye, can detect a temperature difference of .00018–.00036 degrees Fahrenheit between prey and the surrounding temperature. At six inches away, for example, a live mouse can emit fully twice as much heat as a typical background. These heat-sensing pits allow the snake to simply wait for a bird or small mammal to come near, instead of wasting its energy in the search. The snake can first sense the heat and then lunge to strike—all within 35 milliseconds—and then even track the prey by heat if it gets away temporarily. Heat-sensing snakes include saw-scaled vipers, tiger snakes, boa constrictors, pythons, and anacondas. Snakes that also use heat tracking include pythons and pit vipers—rattlesnakes, water moccasins, copperheads, bushmasters, and fer-de-lances.

Another killing method, the venom trick, is used by the viper and cobra families, and very effectively, in some of their 650 species. Venom is the ultimate saliva, or digestive juice (from which it evolved). There are three deadly types: neurotoxic, hematoxic, and a combined form. A neurotoxic venom, the cobra's, for example, attacks the nervous system. The victim's blood pressure and brain's electrical activity drop, and the hapless creature stops breathing at about the same time that its heart stops beating, both paralyzed. A hematoxic venom, like that of the rattlesnake and others of the viper family, acts by breaking down blood cells so that they can no longer carry oxygen to the rest of the body. And a combination venom, like that of the Gabon viper, does a little of each. When a snake bites, muscles around its venom sac constrict, sending the venom through a duct and then through hollow or grooved fangs in the front of their upper jaws. A single cobra may hold enough venom in its body to kill ten to fifteen people.

Venom is also, incidentally, a weapon of two species of lizards, almost all spiders, some ants, bees, and wasps, and even three kinds of mammals: the short-tailed shrew, the European water shrew, and their relatives, the solenodons. Many sea creatures use it also.

Though snakes vary in their use of all these deadly tricks, they have some curious things in common. Most of them twitch their tails when aroused (though only the rattlesnake makes a noise with its own body, as opposed to a twitch of skin against dry leaves). All are quite sensitive to smells through their flickering tongues. All are deaf except to ground vibrations felt in their skull bones, and almost all use their eyelidless eyes to detect movement skillfully. They are "subtile" indeed.

Carnivorous Mushrooms

While toadstools don't rise up and grab dogs, or shiitakes slurp up small children, there are at least eleven species of gilled fungi, regular "mushrooms" that eat small underground worms called nematodes; and there are many more very tiny fungi that eat even smaller animals. Murder methods range from the small noose to a special adhesive, to poison and parasitism.

Watch the oyster mushroom, a typical woodland species, in action on living and just-dead trees. It begins its attack with a toxin (one that no one has yet fully analyzed) to knock out the nematode crawling nearby, then sends filaments out to grow right into the hapless worm, digesting it from the inside out. These meals are tiny but high in protein, and they add up. Another mushroom hangs out several underground nooses, each made of only three cells; when a nematode happens to squeeze through one of them, the cells inflate—very quickly, in one-tenth of a second—to strangle the little worm.

These dramas take place under cover of darkness—underground—where most of each "mushroom" lives. There, filaments called hyphae make a network called the mycelium. The part one sees aboveground is only the ephemeral "fruit" of the fungus; it is, however, essential, emerging from its ready root to spread the spores, some 2 billion from a single common mushroom, so that the plant can reproduce. (This is a method of plant propagation much older than the seed; each of these spores is a clone.) The underground network, always holding the murder weapons, can extend for hundreds of feet and live for centuries, which is why mushrooms often seem to appear in the same general area. The innocent bloom above, the guilty hide underneath, in a partnership of crime that results in regeneration.

Those fungal species visible aboveground, only about 5,000 (including the few carnivores), are a distinct minority among fungi. There are actually some 300,000 species of fungi playing out their lives on Earth, living either entirely underground or inside plants or inside or on us, some parasitically and some symbiotically.

Fungi are never Technicolor green. They lack chlorophyll and so cannot "swallow sunshine" to live the way that green plants can. This characteristic does not account for some species' being carnivorous, however, for most fungi actually eat vegetable matter, dead and living. The ones that do eat meat are among those that have evolved in places poor in nitrogen, especially swamps, bogs, and decaying wood. The story comes full circle with this fact: In the same nitrogen-poor locations are also found the plants that are carnivorous; there are a full 450 species of them, flowering beautifully and eating meat fiercely.

At least one facet of the mystery remains to tantalize: With all the worms and insects in the world, why are there not even more carnivorous fungi and carnivorous plants? Why not even more species of carnivorous birds? One can only wonder and ponder what might lie ahead.

Hot Plants

The voodoo lily pumps out its putrid smell with its own metabolic heat engine, warming itself as much as 55 degrees Fahrenheit above the ambient air on its flowering day. The heat is created hormonally, and, in the process, a natural, biological aspirin is created too.

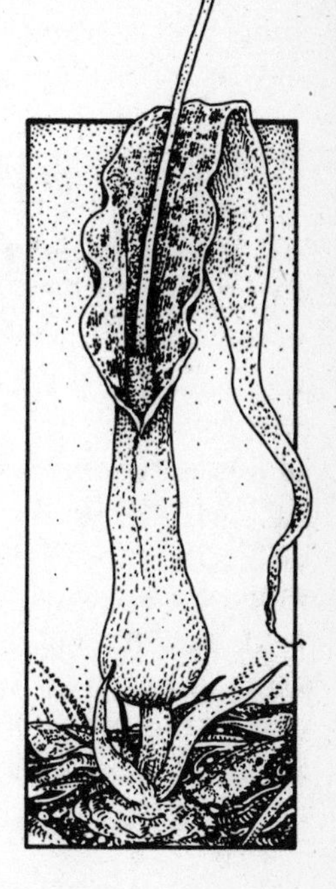

But this bad smelling flower is not alone among plants in creating its own heat, for which many use a metabolic rate as high as a hummingbird's. Arum lilies, the eponymous skunk cabbage (which also uses its heat to melt through the spring snow), some palm trees, and some Dieffenbachia (a common house plant) do so. Even some philodendron flowers warm themselves to attract scarab beetles.

Why would these plants produce heat, especially since most of them live in the tropics? It is, unlovely enough, to mimic rotting meat—carrion—and thereby to attract the insects that love that odor and taste. (Most of these plants are indeed pollinated by carrion beetles and flesh-feeding flies.) At least the heat and smell need only be produced at full bore during the few days of its flowering. As for the chemical aspirin, it is simply a by-product of the heat-creation process.

Piggyback Fish

One species of female angler fish dwarfs her mate and wins the prize for her superiority in size. She weighs 20 to 30 pounds, while he is less than four inches long, sometimes much less. He attaches himself to a female's body after hatching and seems to stop growing, while she, like most fish, never stops at all. Drawing all nourishment from her bloodstream through blood vessels in his head, he is a metaphorical leech. The tiny male freeloader's teeth and intestinal track have mostly atrophied after being along for the ride for so long.

The House of the Chambered Nautilus

The chambered nautilus, known more for its remarkable architectural abilities than for its eating habits, actually enjoys devouring bits of chicken that it sometimes finds in baited fish traps in the ocean. And it enjoys crunching them with a strong parrotlike beak—bones and feathers and all—to add to its regular diet of fish, crab, lobsters, and other sea delicacies.

Swimming slowly through the deep tropical Pacific's coral reefs in a spiral white shell touched with tan brush strokes, the ancient chambered nautilus was the ocean's first large, freely swimming eater of meat. Because of its unusual shell, it was, and is, able to rise from the bottom where its ancestors began and hunt at higher levels, all the way through more than a thousand feet of ocean. Beginning some 450 million years ago, the nautilus and hundreds of fellow species of chambered cephalopods ruled the seas. By the end of the Cretaceous period some 65 million years ago, almost all had disappeared. Their sway gave way to that of the modern bony fishes, faster and more extensive swimmers and better protected against predators, and to the shell-less cephalopods, such as the squid and octopus. The nautilus is the only cephalopod left that has a complete external shell.

And of those that have lived on Earth, only a half dozen or so species of nautilus remain, essentially unchanged in their amazingly adaptive ability to hunt through steep depths of ocean. The trick of such long persistence lies in their architectural chambers. Their first "house" has only about seven of these chambers, and they add some 30 new ones to their spiraling shell as they grow to maturity, always sliding outward to live in the last and largest one, open to the sea, and always sealing off the one behind that with a mother-of-pearl wall. Threading back through all the rooms, like a tightening curl, is a single thread-like organ called the siphuncle. It replaces the fluid inside each "back room," as it is formed, with nitrogen. About a month is required to empty each chamber of its nearly one ounce of liquid. The result is an animal with neutral buoyancy, its weight always in rough balance with the weight of the sea water that it displaces; and since the weight of sea water changes with depth, the creature can move through various ocean levels without having its shell crushed. Its hunting arena is thus larger and more varied than that of a strict bottom dweller.

Living things that last this long can be simple. The chambered nautilus does not seem to be able to add fluid to its chambers, only to subtract it. Its vision from its two large eyes is poor. It cannot move fast, though it does use a form of jet propulsion. And it is notably less intelligent than the squid or octopus. But down in the ocean, ecological conditions do not change as much as they do elsewhere. And so this living submarine has had, in the races of survival, a last, deep laugh.

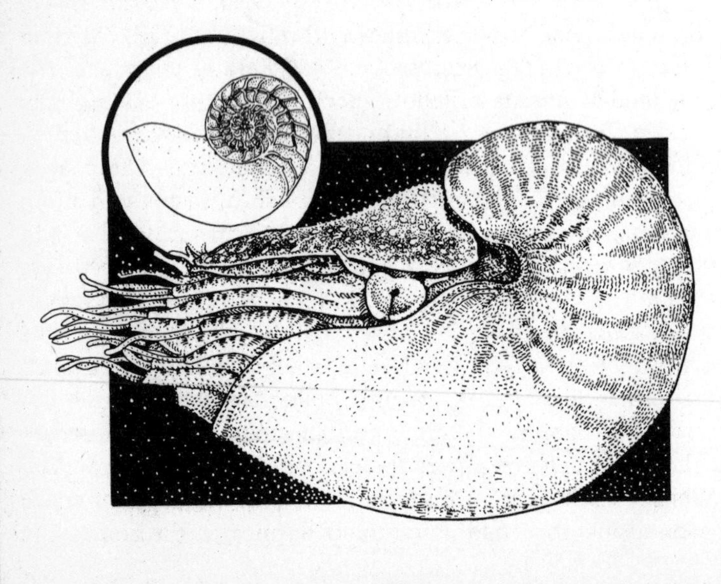

Glowing Sea Dwellers

One shark blazes bright green. A firefly squid is dotted with "running lights." And other lit-up fish have luminous barbs under their chins, glimmering "portholes" along each side, and, in the case of the true flashlight fish, a bright oval of light under each eye. If gathered together, they would look like a neon display for some department store's Christmas specials.

All of these sea creatures live deep down, in dark water, so they need the light. And they use it in a variety of ways, depending upon the species: to see, to attract prey, to frighten away or confuse predators, and to communicate with each other. One squid squirts luminous fluid, from whose glare it glides into safe darkness. Some seaworms wax bright to attract mates. Several fish use their searchlights to lure prey. The flashlight fish use their lights for all these purposes, with the Caribbean species known for darting zigzag to escape from bigger fish while turning its lights on and off.

Called bioluminescence, this living light has evolved, mostly independently, in about three dozen sea creatures, from shark and sponge and clam to protozoan, and it is created in different ways. The flashlight fish proper, for example, wears its light courtesy of colonies of billions of bacteria, all living in a pouch under each eye; these bacteria give off light as a by-product of their metabolism, much as we give off heat as we exercise. The firefly squid manufactures the chemical luciferin as land-dwelling fireflies do. The natural light of these creatures has actually been used in medicine—the ground-up tails of fireflies and the lit-up bacteria of jellyfish can be mixed with calcium and injected into human muscle; there they can make muscle function visible for diagnosis without harming the patient.

Though deep-sea inhabitants are the ones that have evolved the most vivid chemical glitter, bioluminescence is found elsewhere, too. A handful of fungi have light, probably as fallout from their metabolism. And a few thousand insects (out of millions of species) glow and flash, notably the fireflies and the glowworms. New Zealand has a cave called the Glowworm Grotto, where millions of the tiny insects glow in the dark to attract mates—and about 50,000 visitors a year.

Formidable Flowers

A deer with its mouth full of flowers, quiet at the forest's edge in a slice of sunlight; a soft brown bunny, nibbling dandelions: These may not seem like threatening images, but they are to the flowers. The rose family seems to have evolved its thorns—really called prickles—as a protection against deer and other grazing and browsing animals, such as the rabbit:

Both would consume their delicate seeds and thus prevent their reproduction. The tangle of pain usually drives the animals away.

The garden rose is obviously so armed. But so are others in the rose family, which includes Japanese flowering cherry, almond, peach, and apricot trees, as well as blackberry and raspberry bushes, all guarded with some sort of thorn or prickle. Even the wild apple tree, also cousin to the rose, used to have a few well-placed thorns before it was domesticated.

These plants need to use fragrance and color to attract flying insects for pollination. A rose with thorns is thus an exquisite evolutionary compromise. Other flowering plants, taking an alternative strategy, have evolved pollen blasts for the faces of small animals or seeds with indigestible coatings (unlike the rose's) that, when eaten by animals, pass right through them and thus gain a ride to a different location for sprouting.

The armor of the rose family takes different forms. Prickles come in red, yellow, tan, pink, brown, and green, some sharper, some softer, some translucent, and some opaque. They may be straight or hooked, crowded or widely spaced, and pointed up, down, or at right angles to the stem. But none can be defined as true thorns, since they are not modified branches and do snap off cleanly.

The garden rose with thorn, an especially lovely plant in armor, is ancient in its appeal on Earth. A fossil leaf print found from 40 million years ago clearly shows both petals and thorns, and roses are probably 20 million years older than that. A delicate wall drawing from ancient Crete in the Bronze Age suggests that people found them beautiful even then, in spite of the bit of pain that they can bring to many creatures, including us.

Why Are Moths Attracted to Light?

There they are again, the moths, fluttering around a streetlight or porch light or clinging to the kitchen screen. There are probably two plausible explanations for this seemingly pointless behavior. The moth, a nocturnal creature, may have evolved to discover that this light provides a slight bit of protection from its predator, the dark-loving bat. Or, more likely, they may have the light confused with the moon or a bright star, both of which they typically use as a simple reckoning point as they fly at night, keeping that light on one side of them in order to proceed in a straight line. Close objects like electric lights, however, have the opposite effect: Keeping the beacon on one side as it flies, the mindless moth finds itself merely circling it. Pointlessly.

But these mindless moths have at least brought us a lovely poem from Shelley that reads, in part, "the desire of the moth for the star,/Or the night for the morrow,/The devotion to something afar/From the sphere of our sorrow."

The Most Artistic Animal

A great artist, plain in plumage, lives in a place where color in profusion dazzles the eye, a place where food is so plentiful that a gift along these lines would not impress with its calories or color. It is a creature in need of a way to impress the opposite sex, in a species for whom vision is most important. And it has found one: It builds something absolutely lovely.

The artist is the male bowerbird, and the work of art is its bower. This bower is, perhaps, save some of our own species' creating, the most elaborately beautiful structure built in the animal kingdom. Picture an edifice such as this in the jungles of New Guinea or Australia: On a mat of bright green moss, it is a hollow tower of sticks, about two and a half feet high, anchored to the ground and touched with dozens of decorations outside and in. All are bits of color and always replaced as they fade: flower petals, butterfly wings, bits of colored fungus, fruit pieces, bark snatches, snail shells, iridescent beetle skeletons, and bright leaves, along with anything humanmade that can be found (red buttons, matchboxes, sparkling car keys, a plaid sock). The dazzle is arranged by the bird in color zones of reds, oranges, and blues, with some birds using more of one color than others of the same species. Most bowerbirds love blue best, though, and they often save it for inside. Some species even make a paste of chewed up blueberries and, with a piece of bark as a paintbrush, add yet more accents of blue. And this is only one kind of bowerbird. As a final touch, many species add decoration to the path outside, too, or build a kind of garden fence around the bower, also decorated with all the flashes of color. Some make structures like maypoles, others arched tunnels, and more.

The bowerbird swain takes months to build and refresh his masterpiece, trying out a flower in one place, then moving it to another, changing its mind about which butterfly wing should go inside, behaving, in general, a bit like a temperamental artist. Why? To entice females into the bower. Discerning and a bit promiscuous, they will enter, to mate, any bower that they find beautiful. Female choice, once thought unimportant in nature, is now known to be a common key to who mates with whom. It is the females, then, who have shaped this male artistry, bringing it to its glorious level of competition.

Bowerbirds have evolved this sexual strategy instead of bright feathers or melodious song, both common—and usually mutually exclusive—ploys for attention. It is a very unusual one, a lovely small alleyway of evolution. Another bird may borrow delicate spider's webs to hang its nest from twigs (the goldcrest warbler); may fuss over its mound of earth for six months to keep it at exactly the right temperature (the incubator bird); or may even bring in plastic bottles, Sunday newspapers, oar handles, and such to build the nest up every spring (the bald eagle), but no other but the mousey-brown bowerbird creates such beauty.

Undersea Gardens

With names like Venus flower basket, pink vase, and orange ball, sponges are among the flowers of the sea. Far from delicate, though, and definitely not plants, since they eat food particles instead of using sunlight, they are among the world's most successful animals. Sponges have colonized all the seas, at all depths, from shoreline to several hundred feet down (a few live even in fresh water); and many have been as they are now for more than 500 million years.

These creatures come in nearly 10,000 species (with many more, surely, that remain to be discovered) and in sizes from less than an inch long to 100 pounds in weight. Like soft sculptures, they have been molded by their environments into shapes as various as trumpets, trees, vases, domes, fans, bowls, cups, and snowballs. They come in a rainbow of colors too: red, yellow, brown, white, pink, green, blue, and orange. Some sponges are solitary, and some live in melded groups.

These are primitive, many-celled animals, without hearts, brains, stomachs, lungs, nervous systems, or muscles. Sponges do have skeletons, though, a type of tougher tissue that can resemble chalk, crystal, glass, or soft horn, depending upon the species. A natural (not rubber or plastic) sponge found at the drugstore is the hornlike skeleton from which the living tissue has rotted. Natural sponges can be caught by people fishing and can be sea-farmed too.

Sponges function as living sieves, porous processors of sea water. Specialized cells within them sift and consume the food particles found in the flow. Wastes are ejected through one larger hole. Sponges taste bad and so have few predators. They sometimes shelter shrimp, encrust coral, and provide hermit crabs with a living shell covering—all ancient roles, quietly effective.

Are There Any Green Mammals?

Green amphibians, green reptiles, greenish fish, green birds, and surely green trees, of course, crowd the Earth. Yet almost all the mammals that anyone can think of are decorated in golden and reddish brown, various shades of white, gray, brown, and black. Only one mammal can be called greenish—the green monkey—though its fur is really black and yellow.

Why? The reason seems to be that there are two main purposes to color in living things: to hide from other species and to show off to potential mates of a creature's own species. Nonhuman mammals, who don't see color well, don't use it for either purpose. Even all the greenery of the planet is of no particular color value to them, and so they have not taken it on as their own color. Birds, whose vision is superb but who need to be safe, too, have solved the dilemma of safety and show-off by going both ways. Some are very well camouflaged for safety, while others are very brightly colored for allurement. They come, then, in purple, red, pink, orange, yellow, green, and blue, an entire rainbow of color utterly unknown among mammals.

Color in animals is created in three ways—by pigment, structure, or a combination of both. Green pigments decorate dragonflies, sponges, caterpillars, and lobster eggs, among others. Structural colors are created when the surface cells of the animal interfere with the path of sunlight back to our eyes, scattering or refracting it in some way, instead. Green butterflies, frogs, lizards, and snakes can wear their green this way. Structural color acts along with pigment to make some birds, reptiles, and amphibians green, too, but in these cases the pigment is usually yellow or blue, not green, and the structure provides what remains necessary to create the green. Mammals use these painterly tricks, too, but have never ended up with a true green.

Most species of octopus show moods by turning their colors from gray to a brilliant pink if disturbed. Other creatures of the sea use color to send mood signals too, among them the fiddler crabs, which turn purple or another color to fight for a mate, and the cichlids, which have their own entire spectrum of colors.

The sea can be a very colorful place, in general, with creatures who do not change at least as brilliant as those who do. Some, like the blue-ringed octopus (not a quick-change artist), are poisonous, and their coloring helps other creatures to identify and avoid them. In other cases, the bright colors provide camouflage or warnings of territorial possession, necessary in such vivid locations as dazzling reefs. And in still others, the colors enable fellow species members to find each other for mating. "Emotional" color can indicate sexual availability.

Such "sea change" is accomplished by pigmented color cells in the sea creatures' skin. The fish's cells, when governed by nerves, can suffuse color throughout the skin and make it splotchy all in a matter of seconds; and, when controlled by hormones, the color changes take longer to become established but linger longer as well.

The Thrip's Choice

The only known creature that can choose between laying eggs and delivering live young is the thrip, a small (about five millimeters long) tree-dwelling insect whose wings are edged with "hairs" that look like eyelashes. In fact, 12 out of about 4,000 species of thrips have this trick in their repertoire. Thrips laying eggs are rewarded with female young, those bearing their offspring live are blessed with male babies, and a single thrip can switch back and forth. Thrips, probably a 30-million-year-old species, could be a boon to scientists trying to sort out the factors that led to the evolution of the two modes of reproduction.

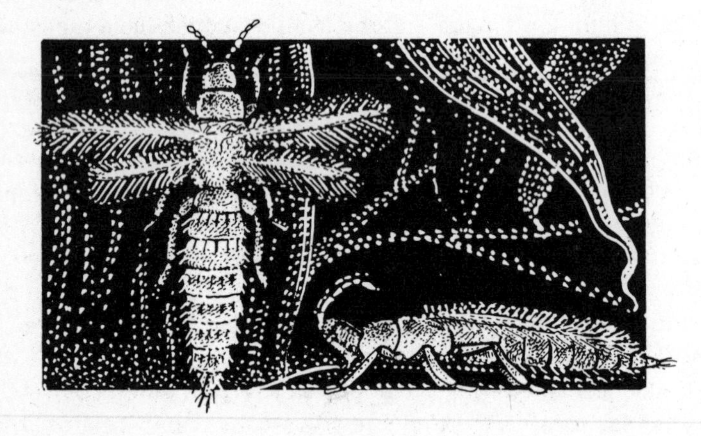

A Home in a Bubble

To contemplate the tiny appendicularian is to feel joy at the everyday miracles of life. It swims, in infinite numbers, through every ocean of the Earth, over all the past 450 million years, in a uniquely fitted and almost transparent bubble of mucus. Each creature, itself so small and transparent as to be almost invisible, is exquisitely equipped with a primitive spinal column and nervous system, a tail, a mouth, a stomach, an intestine, an anus, gill slits, an endostyle (which spins out the mucus), a pharynx, a reproductive system (usually including both male and female organs and sometimes performing both functions from the same organ at different times), a heart, thin blood, and a trunk. All is on a tiny scale, with the trunk, for example, only about five millimeters long. Yet the appendicularian can swim quite well, propelling its "giant" home—a balloon almost the size of a walnut—as it flips its tiny zigzag of a tail.

If the animal's body is amazing, its diaphanous home is only more so. It is a feeding system utterly unique among animals, one that has been compared to human architecture in its complexity. First, consider the exterior. Like tiny Christmas tree ornaments, the outer membrane, usually transparent, can be tinged in yellow, red, or violet and shaped a bit more like an apple or an umbrella or a rounded torpedo. It lies alongside, over, or all around its owner. All this depends upon which of the seventy species of appendicularian one is describing. The membrane gives appendicularians their other name—tunicates—because it resembles a tuniclike garment.

Inside, every bubble is equipped with intricate filters shaped like stacked wings, through which the animal pumps ocean water by flicking its tail. There are also tiny passageways and unimaginably delicate mesh nets, all connected to its trunk, through which it sips food up to its mouth as through a straw. Thus is concentrated some of the smallest foods anywhere: dwarf phytoplankton, whose size is measured in the thousandths of a millimeter, and even bacteria. Some of the larger species of appendicularians are able to remove all of the dwarf plankton from several hundred milliliters of ocean in one hour. Thus it is possible to glimpse their influence on the sea.

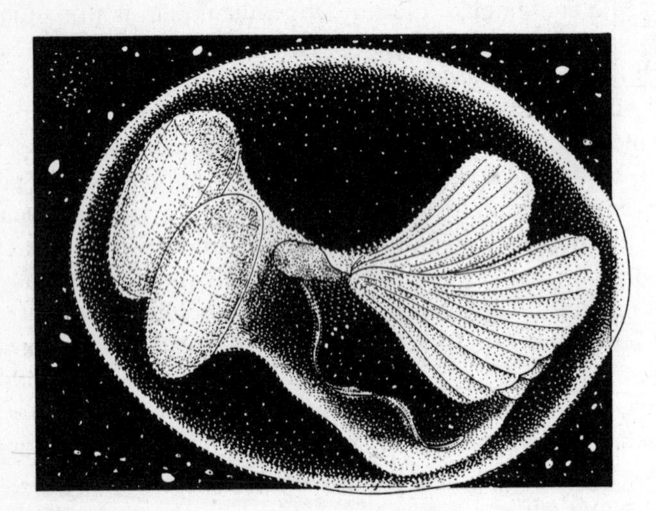

Breeding without Males

Several species of whiptail lizards include no males at all, and not only do the females reproduce by themselves but they sometimes behave like males, mounting each other and pressing their reproductive areas against each other. This behavior, like that in many species in which only the males behave this way, seems to synchronize the reproductive physiologies of both sexes before mating.

This "virgin birth" method, called parthenogenesis, has become the rule for 15 of the 45 species of whiptails. Each of the 15 has developed it through hybridization of closely related species. It is, essentially, cloning, since the females and their offspring are genetically identical.

Study of these unusual lizards has shown that they can reproduce with females only because their brains have retained the pliancy of their ancestors' brains, which had dual circuits for male and female behavior and hormone receptors that were flexible.

Attack by Silk

The silken lace of a spider's web may catch the dew first, then glitter exquisitely in the sun of a spring morning. But this trap of silk, unique among living things and some 385 million years old, is as strong and effective as it is beautiful. Perhaps only four-millionths of an inch thick, it is elastic enough to stretch to twice its original length and can hold five times more weight than a piece of steel of the same diameter. The thread is spun out of the spider's abdomen from organs called spinnerets, and its silk is used by different species to create their webs, chambers, burrows, traps and trap doors, a path through the woods for a mate to follow, egg sacs, nurseries, and courtship bowers, to wrap prey, and even to fly.

Among the nearly 30,000 species of spiders, there is much variety in web-building, though many species (the running hunters) rarely make webs at all. Almost all web-builders, though, begin with a dragline, spun out to make the framework of the web and carried with them wherever they go. This is the paintbrush with which they create their marvel. A typical garden spider uses this line to ride the silk first, beginning a triangle between two branches or leaves, then dropping down through midair to find the third point. It next builds the outer periphery of its web, by connecting several more branch or leaf points. Then the center is built, with several spokes connecting it to the periphery. (Elaborate centers are the mystery of spiderdom: Are they shields to conceal? Engineering reinforcements? Warning devices for birds? Ways to catch water? Something else?) Finally, the concentric irregular circles are spun within the skeletal structure of the web, continuously from the center out and across all the spokes. This is the generic method that each species adapts uniquely. One tiny red spider makes a web like a triangular sheet, instead. The purse web spider builds its in a slight ground depression, and the orb webbuilder spider of Mexico cooperates with 5 to 150 "colleagues" to connect webs together into a giant three-dimensional space.

Whatever the kind of web, it is the spider's "eyes" for attack, since it enables the creature to use its exquisite sense of touch. The average spider can usually distinguish mates from leaves falling onto the web and both from different kinds of insects thrashing, simply by the way the web feels with them in it. Thus informed, it rushes to the site and wraps the hapless insect in silk, sometimes biting it first.

The silk creating begins when the spiders are very young. Just hatched and trailing draglines behind them, they make a tangle on the ground, then soon prepare to escape the family chaos. Reaching the top of a weed, bush, or fence, they face the wind and let it bear their silk away, their bodies attached, until the line snaps from its original mooring and floats down with them into a new place. This flight is called ballooning, and, in a good wind, it can bear a tiny spider tens of thousands of feet high for a while, even, sometimes, allowing it to colonize a new island.

First failures often mean twists of loose silk flying through the air—the word for it, "gossamer," is derived from "goose summer," probably because of the spider's soft down.

Many insects make traps, though not of silk, and many make silk, though not for traps. The spider's architecture for attack is unique.

Medicinal Plants

That a flower could lift a fever may seem like magic. Yet flowers, barks, seeds, leaves, herbs, roots, and other plant parts were our original pharmacopoeia. Not all of anybody's ancestors could have made it without them. Early people had to learn by trial and error what to use to save the sick and wounded. Some of these efforts were based on the plant's iconic properties, with heart-shaped leaves tried for heart problems, and red flowers for blood disorders, for example. A folklore developed and was refined, its mythology lovely still in Shakespeare's time, when he wrote in *Hamlet,* "There's rosemary, that's for remembrance; . . . and there is pansies, that's for thoughts . . . There's fennel for you, and columbines; there's rue for you . . ." By as early as 5,000 B.C. the Chinese had amassed a full shelf of powerful plant medicines, many of which worked beautifully.

Even today, the best estimate is that up to half of the prescriptions written by American physicians contain a drug derived from a plant. A few examples: Digitalis, the widely used drug for congestive heart failure, was developed from the purple foxglove; cortisone is derived from a chemical found in yams; atropine, a drug for stomach ulcers and for dilation of the eyes, comes from the belladonna and other plants; reserpine, used to control high blood pressure and as a tranquilizer, was developed from snakeroot; commonly used antileukemia drugs are derived from the periwinkle; morphine comes from poppies; the muscle relaxant tubocurarine came from the curare vine; and most antibiotics come from bacteria and fungi.

Helpful plants span a broad spectrum too. The bark of the willow and birch trees yields a natural aspirin. The seeds of the Ammi plant can be made into a drug to relieve asthma. Many of the root "chewing sticks," long used to clean teeth in Africa and elsewhere, are now known to contain natural antibiotics, anticavity ingredients, and fluoride (some even combat sickle-cell anemia). Both quinine, derived from a bark, and tea brewed from the neem tree's leaves help against malaria. Ipecac, a root common in Central and South America, is used to treat dysentery and to induce vomiting. Chenopodium seeds fight intestinal worms. The nasal decongestant ephedrine comes from a Chinese shrub. Camphor and menthol are also plant products. And poisonous ergot can also yield a medicine to hasten labor in problem childbirths and even relieve migraine headaches.

It is, after all, only reasonable to expect that plants—ready-made chemical structures that have evolved on the planet—should create effects on the Earth's animals, including people. The most biodynamic chemicals within plants seem to be their alkaloids, glycosides, essential oils, fatty oils, resins, mucilages, tannins, and gums.

Our modern plant-medicine shelf probably holds only a few of the medically active plants on Earth. Of the some 300,000 species of flowering and nonflowering plants on Earth, very few have yet been tested—including many of those now claimed to work in folk medicines—though a small program is now going on to collect and test some of those in Amazonia. Any death of flowers or other plants by extinction diminishes this beautiful potential for health. And diminishes the world in many ways too.

The Furriest Animal Alive

To be absolutely sure, pat one. Musk oxen, big and shaggy as unkempt buffalo, are the furriest animals alive today. An adult's fur may be easily six inches deep. The animals are outfitted for the Arctic in brown, tangled guard hairs and a lighter silky underhair called *qiviut;* the *qiviut,* eight times warmer than wool, can be knitted into soft sweaters, scarves, and hats.

These creatures are neither precisely oxen nor musk. They are a relative of wild cattle and of goats, though they look more like water buffalo, with curved horns and a beard. Native people call them *oomingmak,* or "bearded one." Herd animals, they trundle in groups of eight to fifteen along low-lying plains and

river valleys where vegetation is relatively ample, gathering in a circle with their heads out if it becomes necessary to defend themselves against Arctic wolves. The young are tucked in the circle's center, protected by adults 600 to 700 pounds heavy. A furry success during the Ice Ages, they ranged, then, all over Europe, scraping snow crusts off of the mosses and, if necessary, dropping their heavy heads down to break a sheet of ice. They see well, even by the light of the moon and stars in their now Arctic kingdom.

An Artichoke on Legs

It eats ants and termites—about 200,000 a day—but may not be related to the anteater. It has a foot-long tongue that darts out and probes like a length of sticky flypaper, then folds up for storage in a pocket in its throat. It has good claws

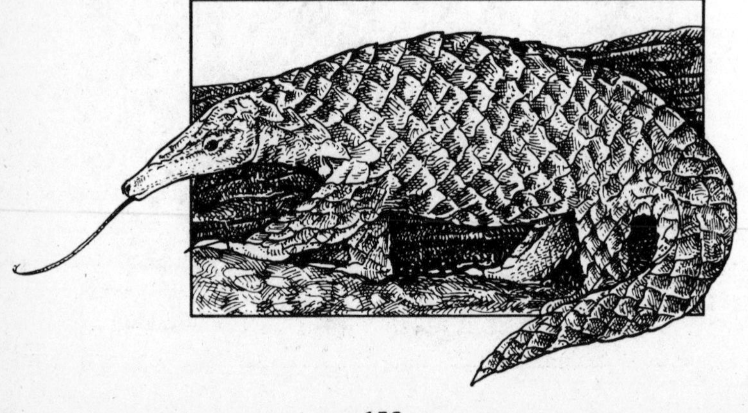

for digging and no teeth. Its body is shingled with tough, sharp scales that can open and shut quickly, a body covering unique among mammals. When threatened, it can lash out with its tail, spray out a smell worse than a skunk's, or roll up into an armored ball. It sleeps all rolled up too. As a species, it may well be as old as the last of the dinosaurs. It is the pangolin.

This odd, well designed, usually nocturnal creature lives in Africa and Asia, where its seven species range in size from a house cat to a large dog and hardly make a sound. No one knows its life span, but naturalists worry about its survival, since its scales and hide are considered medicinal, lucky, and good for boots.

Winter Survival

Winter throws down an ultimate, desperate glove of challenge to living things, and one that, like all of our own gloves, comes trimmed in ice every year. Birds must survive blizzards and freezing days, fish their frozen lakes, insects the cold snaps, plants the shrinkage of sunshine, and all animals the snows. Food disappears, whether it is berries, grains, flowers, smaller creatures, or sunshine itself, while cold increases warm-blooded creatures' need for food. The coldest winter weather can even freeze cells: As the water inside the cells expands to form ice, its stretching, piercing ice crystals can break cell walls and membranes. Also, ice between cells can shrink them or sometimes simply turn them to mush suddenly as the ice thaws. The result can easily be death. Trees can even "explode" in a mighty crack, much as the asphalt on roads cracks from a freeze/thaw cycle. Thoreau called it "the iron age of the year." It is, he said, the season of "perfect works, of hard twins

. . . The leaves have made their wood . . . able to survive the cold."

Yet, if the northern world is to be habitable, winter must be survived one way or another. There are three choices: hibernate, migrate, or, in this case, cope. All three are risky—winter winnows the numbers of living things every year. But across the whole span of living things—plants, birds, insects, and other animals—the most common of the strategies is to cope.

This choice includes a dazzling array of tricks, from simple huddling to self-insulation to mini-migration, mini-hibernation, even strange changes in body cells. They add up to the kind of miracle that can be engendered only by evolution, the drive of life to hold the northern world as part of its domain, a place where spring, summer, and fall are magnificent rewards of plenty. And even where being there first in the spring means, for birds, getting prime choice of territory.

The first choice is huddling, which sounds like a simple strategy but often works quite well. Birds are very good at it, first fattening and growing more downy underfeathers, then fluffing up into a ball. Some, like a covey of quail, form a circle to make a kind of common nest out of their wings. Ptarmigan huddle under the snow sometimes, too. And commensal birds, such as the sparrow, will huddle up against the lights of theater marquees and other humanly engineered oases of warmth. Musk oxen and buffalo huddle, too, in their herds against a blizzard. Some small mammals, like flying squirrels, huddle in their homes together on especially cold days. Shivering is a way all creatures warm themselves, but one kind of owlet moth, a winter noctuid, can begin shivering at very low temperatures, enabling it to make winter into its main season of life.

Changing one's growth pattern—either by stunting or enhancing growth during wintertime—or having evolved to be tiny or huge to begin with, works well in winter, too. Life on the Arctic tundra, where winter cold lasts much of the year, is noticeably low down: Clinging mosses, lichens, and shrubs all hug the ground. In places like Iceland, what trees there are, are all small (the dwarf birches are beautiful), and in Alaska there are areas of forest all in miniature evergreens. High in the snowy Andes, one plant lies low enough to be called living snow. Rates of growth tend to be also very slow in the North. Reindeer moss spreads less than one-eighth of an inch every year, and the Siberian larch grows less than an inch each year.

The reasons for these northern patterns of growth and the tiny kingdoms they create are several. Sunlight, for example, which hits the ground only to be re-radiated upward, is thus usually its warmest close to the ground. And shallow root systems, which prevent prolific growth, are a northern necessity, since the soil is often poor or partly frozen. (In the less extreme North, the growth of trees only seems to stop in wintertime. Actually their roots keep growing.)

The opposite growth trick for wintertime, enhanced growth—a lifetime hugeness—exists too. A good many living things in the northern world are simply behemoths. Think of musk oxen, mammoths, giant ravens, and huge evergreens that grow large for heat retention and also, sometimes, because of less competition for resources in the North. Creatures of the northern seas are also large. In a cold sea there is no place to hide or huddle in the all-encompassing cold, and a larger sea creature is thicker and stronger. It also lives longer, enabling it to survive a very bad year or two and still reproduce. In the North live giant sea urchins, a sea anemone spanning two or three feet, sea stars with 24 rays, sea slugs so big you wouldn't want to meet one, a clam that weighs up to ten pounds, and the giant octopus, all found off the shore of the Canadian province of British Columbia alone. And there are the whales of the North, too, warm in all the cold seas of the world.

Insulating oneself often accompanies these strategies. The squirrel, the whale, and the deer put on fat, the redpoll and chickadee grow more feathers, grouse develop fringed feet as natural snowshoes, and many creatures shiver successfully, which generates body heat. Those with tails or feet that stick out gather themselves into balls, sometimes also constricting the veins closest to their skin to avoid draining surface blood of its heat. Small mammals, such as the shrew, are so expert at insulation that they actually gain weight in the winter. They take advantage of "brown fat," a special kind of fat stored over their shoulder blades that is especially warm in its "sweater" position and also when broken down for heat energy. Perhaps the best insulation of all, though, belongs to the seeds of the far North: a protective envelope *par excellence.* Frozen seeds of the lupine bush found buried in the Yukon tundra and estimated to be 10,000 years old have been uncovered and found still viable. So also have oriental lotus seeds, found buried in a Neolithic canoe in a deep peat bog. Placed in warm water, they promptly sprouted.

Color can occasionally be a winter trick too. It is a protective camouflage for the Arctic fox and wolf and the weasel that turns white only for wintertime. But it can also be more. A white polar bear hardly needs its white for protection, and the polar bear color may even seem to deflect sunshine, but it may have another function. With less of the darkening pigment melanin taking up space in the hairs, more air space is available to trap the bear's own body heat. And it is also possible that the white fur reflects heat from the animal's body back onto itself. The green of evergreens also enables the trees to avoid the metabolic expense of creating a new supply of chlorophyll every spring.

Two other tricks are mini-migration and mini-hibernation, which is called dormancy in plants and many animals, and diapause in insects. It is truly minimal. Some insects move down from their trees into the grass, while other insects and some fish drop to a lower level in the water. All fish, beneath their gray speckled ceiling of ice or cold water, and most birds that overwinter, enter torpor to some degree, at least at night, by lowering their body temperature to conserve their supply of stored energy. In daytime, under extreme conditions, the opposite sometimes happens, though. In the high Arctic winter the common redpoll can survive periods of minus 40 degrees Fahrenheit by temporarily increasing its metabolic rate to three or four times normal.

Since cellular life and death is the bottom line, there are many exotic strategems of this type for surviving the winter. Evergreens harden and transfer water out of their cells to keep them from ice damage, enabling species such as the red saxifrage to live even within a few hundred miles of the North Pole itself. Wintergreen leaves store sugar and an elixir of other chemicals to create their own antifreeze. Known as supercooling, this inner winter armor can also be built by the sugar maple, red oak, shagbark hickory, and others. It too is effective down to minus 40 degrees Fahrenheit. Trees of the higher North, such as the tundra birch, spruce, and pine, can even accept some ice in their tissues without incurring significant damage. Many animals produce glycerol as antifreeze to lower the freezing point of their bodily fluids; among them are the glacier midge in Nepal, ants all over the northern world, the primitive coelacanth fish, and the mawson, a cod found in Antarctic waters. This antifreeze has been studied carefully in the winter flounder and found to interfere electrically with the formation of ice crystals. Yet other cold-water ocean fish have a different kind of antifreeze, one that prevents the growth of ice crystals even within "freezing" body fluids. And the icefish has no red blood cells at all, only antifreeze. Other creatures, notably insects and fish such as the Alaskan blackfish, employ the cellular opposite of antifreeze: They let themselves freeze virtually solid, allowing ice crystals to fill their guts, blood, and spaces between cells—everything but the cells themselves. They do it by manufacturing various cryoprotectants, chemicals that can reduce the amount of water inside the cells. The blackfish can actually survive several months this way.

Of course, these cellular strategies can cut down on a creature's activity. Many spend a good part of their lives completely out of commission. The painted turtle can safely allow itself to be about 54 percent frozen. One type of frog is frozen hard as a rock all winter, then leaping after insects come spring. The prize goes to the Arctic ground squirrel, which can survive a 27-degree-Fahrenheit body temperature regularly. Surviving thus, partly or wholly frozen, is now intractably impossible for another species, people, but there are indeed scientists working on it.

A sweeter and simpler winter strategy is basking—orienting the body to collect maximum sunlight whenever it can. When it warms up in wintertime, watch the birds and squirrels scurry and then stop in the sun. In the higher North, true basking is a specialty of the Arctic woolly bear caterpillar. And perhaps the strangest scenario of all in water is the small mudminnow, under the ice for four to five months of the year; it breathes gas bubbles all winter long.

Winter survival is a necessity for many living things. The lives of animals and plants from the chipmunk to the Siberian larch would surely be shorter if they did not have a winter torpor. All have evolved to the rhythm of the seasons; the bird that sings, the squirrel that starts its family, the flowers that bloom. Apple trees, for example, require close to 1,000 hours of temperatures below 45 degrees Fahrenheit in order to make their blossoms sweet and sure. Without winter there would be no spring.

The Real Roadrunner

Running along at 20 miles per hour, the roadrunner does not go "beep beep"! (It goes "clack," "coo," whines, and makes a dozen other sounds.) In fact, it is a bird beautifully adapted to life in the desert. Feisty, it can take on a rattlesnake as a prospective meal and sometimes win by pecking the snake's head hard enough. Flexible, it eats plenty of insects (also tarantulas and black widow spiders), mice, bats, snails, lizards, even pieces of plants. Well-calibrated, it can reduce its body temperature by 7 degrees to conserve energy during the colder desert nights—then spread its wings in the morning, exposing a patch of skin on its back that collects enough sunlight to raise its temperature again. Small at only nine inches tall, it can run its 20 miles per hour on strong legs. The roadrunner does not choose only highways for running but also speeds along deer trails and other clear-track places where insects are likely to be exposed and common. A speedy creature.

Animal Altruism

"Nature is red in tooth and claw," the old expression went. Animals were thought always to compete fiercely within their own species, prey upon other species, or use each other in various forms of parasitism. Some of this red must now be accompanied by what looks suspiciously like a heart of gold among some animals. In the animal kingdom, helping others of the same species goes well beyond the care of parents for their offspring, the clearest case of extending one's own genetic inheritance to the future, and so a primitive, powerful, and not very surprising thing. Many immature birds help their own parents and other nesting pairs to care for the next generation. Worker bees never do produce young, but work and even die for hive mates. From plasmids to people, organisms help each other. Even dinosaurs helped each other with their young.

At first blush, this kind of behavior—called altruism—is a puzzle: How does it benefit the individual animal who does the helping? How can such a behavior ever have evolved to begin with, since it seems as though the helpers, by not breeding as fast or not breeding at all, cannot be passing along their genes (for helpful behavior or anything else) to future generations? The answer seems to be that helping often does benefit the individual ultimately, or enables its close relatives to pass on many of its genes, or somehow encourages other animals to reciprocate later. In fact, altruism is sometimes seen among birds raised near each other in a colony, even though they are not related.

Altruism is especially pervasive among birds and social insects. More than 100 species of birds, including blue jays and sparrows, are now known to exhibit it. Green woodhoopoe birds, an African species that lives in clans, present an especially good example. Among them, young birds already capable of breeding instead bring food to other nesting females, then feed the nestlings and fledglings, all the while also helping to guard the breeding pair's nest. Though all the birds in the troop are probably at least very distantly related, these helpers are usually, but not always, the aunts and uncles, older brothers and sisters, or even the half-brothers and sisters, of the new infant birds. Their behavior has nevertheless been clearly shown to help the helpers themselves, thereby providing probably the most plausible rationale for its evolution. By assisting for a while, the helpers place themselves in an excellent position to colonize new breeding territory later as it becomes available—which it does only infrequently. They can safely move into it because they have developed allies among some of the birds they helped to raise, who also become their later nest assistants. And they enter their breeding role with valuable practice in rearing young birds. Those that do not colonize new territories may mate with one of the original breeding pair that they helped if that bird becomes widowed, as many do each year. The helpers become so well set up for long-term reproduction that they have been shown to leave as many offspring over their lifetimes as the original nesting pair does.

More drastic altruism is present in many social insects, whole subgroups of which will even give their lives for their fellows without ever breeding at all. Worker bees are a prime example. Some sting intruders—which is suicide for bees—to protect the hive. Others remain sterile, spending their lives feeding hive offspring not their own. In fact, quite a few species of bees, ants, wasps, and termites are considered fully social insects—those that by definition demonstrate cooperation, caste specialization, and altruism. The reason seems to come down again, but very clearly, in this case, to passing on one's genes; this is not only the most powerful goal but one that also carries with it at least the possibility of building a strong species over the long run. In these ancient insect species, there has indeed been time for the individual and the social payoff to converge.

What got the system working among the social insects was probably this: In most of their species, one of the sexes is born from unfertilized eggs and so has only one parent and one set of genes. A female then has a close genetic tie to her full sisters, one that seems to obviate the need to bear young. Staying with her mother to help care for her newest sisters becomes the best way to maximize the future chances for her genes. Insects follow this strategy without, scientists think, always being consciously able to recognize their sisters.

Among even lower organisms, a type of altruism can be found. A viruslike particle—called a plasmid—that lives within a bacterium will occasionally concoct a toxin that kills it, its own host bacterium, and the bacteria around it that do not include other plasmids (the other plasmids guard themselves with a resistance to the toxin of their fellows). The original plasmid commits this kind of suicide when bacteria become quite crowded around it, thereby creating more room for bacteria that contain plasmids.

In higher organisms, especially mammals, individuals who are altruistic beyond the circle of their own young tend to adopt longer-range altruistic strategies. One baboon may help another in a struggle for a mate, yet not benefit in this breeding triangle; instead, it can expect to receive the same kind of help itself later on. Called reciprocal altruism, this differs from the "kin selection" strategy of the social insects (which help close relatives) and the purer "Darwinian fitness" strategy, in which the goal of benefiting one's own genes is the only factor that is immediately clear. Species can, of course, be altruistic in more than one of these three ways. Human beings can show all of them, though they do not always do so.

Perhaps the animals that have not developed altruism, or those humans whose altruism has been damaged by cultural factors, are the poorer for it, even individually; that is, pure selfishness, even that beyond the walls of the nuclear family, may not be adaptive after all. Or altruism may be the highest, purest selfishness that transcends selfishness.

As the naturalist Joseph Wood Krutch said of all animals, "To be an animal is to be capable of ingenuity and of joy; of achieving beauty and demonstrating affection. These are surely not small things, though there is danger that we are forgetting how far from small they are."

"Watchdog" Bird

In the Middle Ages, many families had an area under the stone stairs that was known as a goose hole. In it lived the family watchgoose, who protected the house by hissing, pecking, and wing-attacking any intruders.

The hiss of a goose, much like the bark of a dog or the rattle of a rattlesnake, is a threat gesture. Among geese, it is one stage of ritualized fighting, an attempt to see which will attack and which will back down, all without resorting to violence. Usually emitted with the neck extended forward, the hiss is usually a compromise between aggression and fear.

The Oldest Living Earth Dwellers

They are not an extra troop of wrinkled elephants, a pod of huge blue whales, sturgeon, a flock of odd birds or insects, or even a dinosaur family in days gone by. Though great size, along with a slow metabolism, usually accompanies long life among animals, this group of living things does not follow that rough rule . . . and some of them may even be said not to know death at all. They are the plants.

The oldest known individual plant is about 5,000 years old. It is a bristlecone pine, part dried out and all gnarled—and not large—clinging to a cliff in peace in California's White Mountains. There, it has managed to avoid all the common causes of death among trees: lightning's slash or fire, insect attacks, fungal rot, bacteria overload, or human interference. Its species also grows slowly and has a great deal of resin to protect the wood from excessive moisture, which would lead to rot. If this tree, named Methuselah by scientists, possessed a memory, it could tell us of the building of the Egyptian pyramids, the births of Socrates and Jesus, Renaissance literature, and all the wars, glorious architecture, and Olympic foot races since.

Other trees with dramatic life spans include sequoias (about 3,500 to 4,000 years), cypresses and yews (2,000 years), oaks (1,200 years), and spruces (1,000 years). All dates are based on core sampling of the tree's rings, with adjustments made and dates checked by noting evidence (in the tree and elsewhere) of key factors that affect growth and thus the rings: seasonal wet and dry periods, longer climate variations, large volcanic eruptions (which cause temperature differences), even "earthquake scars" seen when tree roots are disturbed or the whole environment is changed by a major quake.

These trees are not contenders for immortality, however—they will all eventually die—but some other plants truly are. There is a box huckleberry bush that may well be 13,000 years old; a creosote bush at about 11,700; and there are various patches of buffalo grass, prairie bluestem, cattails, and even fairy-ring mushrooms (most of which, like all mushrooms, lie underground), as well as bulb plants such as daffodils, that may well date back to the end of the last glacier, about 10,000 years ago in northern climes, or even much, much longer in tropical areas.

These are not individuals, however, but clones. The plant we see today is probably genetically identical to the original plant that lived back in ancient history, because it is by such cloning that these plants reproduce and have reproduced in one general area.

The easiest to picture is the creosote bush, denizen of deserts in the southwestern United States and Mexico. It starts from a single seed, then splits off its lower stems as it grows. These stems then put down new roots and the cycle continues. The bush's clones are farther and farther away from the now-empty center of the patch where the original generations grew, but they form a rough ellipse so that one knows whose genetics to check. The most extensive ring, called King Clone in California, is 70 feet across at its widest place. A box huckleberry that is striving for an ever greater prize is spread over 100 acres, all, again, genetically the same plant.

There are no tree rings to count here, of course, and scientists must estimate age by how much these bushes have spread, figuring in their average annual growth. In the case of the long-lived grasses and bulbs there is no way to estimate at all, only to say that patches of them could conceivably be the oldest of all. Or perhaps it is the individual bristlecone pine that, all by itself, deserves the prize for longevity.

Can Ants Really Devour a Human Being?

South American army ants and African driver ants—a double billing of horror—typically march through the jungle in a search party of 150,000 or a full army of about 2 million ants. It must be an impressive sight to watch them grab, bite viciously, and then eat anything in their path. (They move so often because they eat up everything so quickly.) Be glad that you are not a spider, beetle, grasshopper, cockroach, lizard, an immobile python, a tethered horse, or, we hope, a baby in a crib or a wounded or sick adult—all could easily be consumed by the "army." Healthy adults—unless shocked while asleep by thousands of bites at once (an unusual situation)—can easily outrun them. Giant anteaters do so regularly, thus according these ants considerable respect.

The ants get their orders from chemical signals called pheromones. The pheromone that stimulates the army to move and keep moving for weeks on end is produced by the ants' own larvae, the littlest ones of the army, carried along. Fanning out first, to look for prey, are parallel columns of ants, six to eight soldier ants across. Then behind them, in wider columns guarded at their flanks by more soldiers, come the worker ants, many of these carrying the larvae. When prey is discovered, the message is spread quickly by another pheromone, and the ants swarm, some crawling into the animal's nose, throat, and eyes, and every ant biting and finally cutting it apart. Even an elephant panics when they crawl into its trunk. A person whose jungle house is attacked quickly goes somewhere else for the night but can return the next bright morning to a truly pest-free home. When the larvae are ready to begin the next stage of their lives, as pupae, they change the signal, causing the army to bivouac. The ants crawl around a tree or large stone in a vast cluster, making a large living ball of ants. Through this clinging mass, passageways are left for the queen and the pupae inside.

Besides running away, there is only one protection against these ants: fire. Pour gasoline on the ground and light it, or hurl flamethrowers, all quite persistently, to deflect the army. It might also be wise to grab the running shoes and the baby.

Lemmings and Suicide

Lemmings—small brownish northern rodents—have traditionally been pictured poised en masse at the edge of a craggy cliff by the sea, near what looks like Hamlet's castle of indecision in Denmark. Then, all of a sudden, by the millions, they are supposed to leap into the icy ocean, a few swimming frantically for another angst-filled minute or two, then cold suicide for all. In an earlier scenario, more folktale than grade B movie, they are supposed to die from pelting by heavy rains, then reappear by tumbling from the sky during the next spring's storms. A new and a Nordic version of "it's raining cats and dogs."

But it is safe to say that any species prone to behavior this bizarre would not have survived long enough for these tales to be formulated. Darwinian natural selection, which works by rewarding the genes of individuals who survive to reproduce—and not species who clear the decks for other troops even of their own kind—would have winnowed out most creatures like this millions of years ago. And all this begs another point: No animal of this level of intelligence—perhaps no animal at all save humans—is mentally complex enough to commit suicide.

But, as with many popular myths, there is an element of truth to the drama. Populations of lemmings do indeed rise and fall, and quite drastically, often in three- or four-year cycles. A field of just an acre, filled with one hundred of them, all chomping at plants and digging at their burrows, can be home sweet home a couple of years later to only one or two lonely lemmings.

One hypothesis explains lemming cycles as the result of many factors acting together. Lemming populations boom as they feed on abundant nutritious plants with few toxins. As numbers rise, the lemmings disperse, leading to frequent aggressive encounters between nonrelatives. The stress leaves the lemmings less able to breed and perhaps injured or sick. Meanwhile, the plants offer less nutrition, having flowered and been heavily grazed. Thus, the now-peaking lemming population is faced with depleted food and the effects of stress. The bust comes as weakened lemmings are taken by predators, and perhaps diseases, that respond to their high numbers and weakened condition. The next boom will begin when predators have left and the plants have recovered.

This population crash and crush is actually a quite common denouement. Population numbers of many small rodents, including field mice and voles, behave in a similar way. And, even less obviously because their life spans may be longer, so do populations of creatures from gypsy moths to lynxes. Wild swings of animal population numbers may sometimes be explained by the theory of chaos, a relatively new area of science and one that is proving immensely useful in this and other fields. It demonstrates that small but important differences in initial conditions can unfold quickly into large and dramatic effects. This science of runaway conditions seems to suit, intuitively, the scurrying, furry little lemmings—whose behavior can be explained by little else.

The King of Insect Migration

On small, strong wings they flutter by, black and luminous orange, the monarchs on their way to Mexico. By the time they settle quietly, covering whole mountain trees with their petal wings, some will have flown more than 2,000 miles. None of the butterflies has made this trip before; instead of experience, their guide seems to be a hormonal sensitivity to the changing day length in the North and an inner magnetic material that enables them to orient to the Earth's magnetic lines of force for direction.

It is not only the monarchs. Painted Lady butterflies cross the Mediterranean in their trip from northern Europe to the northern fringe of the Sahara. The Bath White butterfly traverses the Pyrenees. Some hawk-moth species cross the ocean. Ladybugs migrate across whole continents. Migration is actually relatively common among long-lived insects such as butterflies, moths, dragonflies, and locusts; in the latter group, it is often overpopulation or unfavorable weather conditions that cause the sporadic moves.

But monarchs are the kings of insect migration. By mid-August they are gathering in southern Canada, the Midwest, the Northeast, and the Northwest, building their strength on the nectar and pollen of red clover. By late September and early October they leave for the South, flying along some seven north–south avenues. Beginning with flocks of hundreds, they build to thousands, then millions, as flocks converge to the South. At a pace of up to 80 miles per day, the butterflies migrate by daylight and usually within 20 feet of the ground. At night they rest in "butterfly trees," usually choosing ones in a small grove on a hill near a landmark. Their destinations are southern California and northwest Mexico for the West Coast monarchs, the Gulf

cks, and mid-Mexico for the midwestern
destination, some 30 million monarchs
necting the mountain trees into a scarf
lies. An anonymous haiku writer ob-
an, too, and wrote, "Fallen petals rise/
tch/Oh . . . butterflies!"

soı oyage enough, but by late February or
xican migrants stir, mate, and begin to
e, then, by about the time they reach
s. Their eggs hatch, develop, unfold
the journey north, also probably mat-
vay. The species is carried north in
ping generations; the butterflies that
reach the northern tier of the United States and southern
Canada are probably the grandchildren—or, in harsh years, the
great-grandchildren—of the butterflies that wintered in Mexico.
Over the summer, in their milkweed mecca of the North, the
July generation mates and feeds at relative leisure over their
six-week lifetime. Their children must migrate in the fall.

The Bird That Eats with a Spoon

One type of bird has adapted so that it finds its food not by vision but by sweeping its bill back and forth. Aptly named the spoonbill, this bird swishes its bill through the water, a bill whose edges are equipped with sensitive nerves

that make it snap when it finds food. It never sees dinner at all—and so can feed all night when necessary. The birds also use their spoons for mating, rubbing them together in seeming ecstasy. They are much more dramatic than the avocet, a shorebird that also sweeps its bill.

A Gulf Coast bird that also lives in the Caribbean and in South America, the roseate spoonbill gets its color from the food it eats (foods rich in carotene). Once hunted mercilessly in America for its beautiful plumage, it now lives mostly in protected sites and is thriving. Its paler relatives, which also transfer colored carotene from their food into their feathers, live in Australia, New Zealand, New Guinea, Africa, and Eurasia, where they need more protection from their less colorful human neighbors.

Feats of Bird Flight

A hummingbird hovers, earning its Brazilian name of "flower-kisser." A sea gull glides from dock to cool ocean waves. A vulture soars over the African plain, its wings still. And the swift, a scissors in flight, flaps its wings fast enough to sustain speeds of 100 miles per hour. These four kinds of bird flight—hovering, gliding, soaring, and flapping—each have a poetry and a physics; all are within the repertoire of almost all birds, though they may differ in their medley of skills. Having evolved the immense advantages of flight, only 46 out of some 10,000 living species of birds have abandoned it.

Even our mythic angels could not possibly fly the way birds can, the angels' thick wings attached awkwardly to heavy bodies; neither could Daedalus nor Icarus of Greek myth, who adorned themselves carefully with feathered wings, only to have the adhesive on Icarus's melt in the sun. The design of a flying creature cannot be wish fulfillment but must take account of aerodynamics as surely as must the engineer of an airplane.

Hovering—the kind of flight used most often by small birds like flycatchers, sparrows, and bullfinches, and by larger fishing birds such as kingfishers and terns, usually for only a few seconds at a time—is a defiance of gravity that requires great energy output. No significant air force helps with the lift, which must be generated by the bird's own wing muscles. The body is held nearly vertically, wings flexed at the "elbow" and moved back and forth quickly in the horizontal plane. A kind of clap, fling, and twisting flip of the wings can help, too, in some species, quite different from the steady aerodynamics of flapping flight. Common, too, among small bats and many insects, including dragonflies, mosquitoes, bees, moths, butterflies, and hover flies, hovering has reached an art in the hummingbird.

With a heart and lung surface three times larger for its size than the pigeon's, it can move backward or straight up, turn on a dime (or a flower), and come to a standstill from full flying speed.

Gliding flight, probably the most primitive kind and practiced by flying squirrels, flying snakes, flying possums, and flying fish, as well as by birds, is easier for creatures who can present a broad shape to the wind. It is simply a slowed, directed fall.

Usually combined with gliding is soaring, practiced by large birds who are too heavy to fly very long on flapping wings. To soar, a bird like a stork, vulture, or cormorant finds a free lift, usually in the form of winds rising as they hit slopes or hot air rising in columns over an area of especially warm ground. The latter current, called a thermal, forms most often over clearings in the woods, farm fields, plains, and even cities. The bird merely gets going, spreads its wings, and lets the air current do the lifting. Since thermals are generally not created until almost midmorning even in hot climates, vultures will hardly fly before then.

Flapping flight, perhaps our most common image of bird flight, is created when the wings act as air foils. The wing slices through air, twisting slightly on the downstroke, forcing the air to travel farther over the curved upper surface of the wing than it travels under it. This difference creates slightly lower air pressure over the wing and slightly higher air pressure under the wing, lifting the bird. Without the slight twist of the wing but with a motor to help, this same principle is what keeps the airplane aloft. Swifts have been clocked at almost 100 miles per hour over a measured two-mile course using this flying method. Also, birds can briefly resemble torpedoes through air or water, smoothing their edges to swim, like a penguin, or dart down for short plummets at 200 miles per hour, like the peregrine falcon.

Besides their usual flight mechanisms, birds are helped to fly by bodies designed for lightness. Their bones are almost as hollow as reeds, and some young birds, like the petrel, must, after growing up, sit in their nests for several days and lose baby fat before they are ready to fly.

Flight is an everyday miracle that has evolved independently in birds, bats, flying insects, and flying lizards. It has surely been adaptive. The few bird species that have lost their ability to fly—such as the ostrich, rhea, emu, and kiwi families—did so as their land predators disappeared and they no longer needed to make the effort to take to the air for safety. With that problem taken care of, life on the ground was apparently satisfactory.

A Toad That Squirts Blood

Do not tangle with the Texas horned toad if you don't want to be squirted with blood. Found throughout western North America, and living underground except when the weather is warm, this lizard is four to five inches long and comes in gray, tan, brownish, reddish, or yellowish colors. Decked out with a crown of spikes on its head, it is squat, eats a lot of ants, and is active when the sun is hot. When the crea-

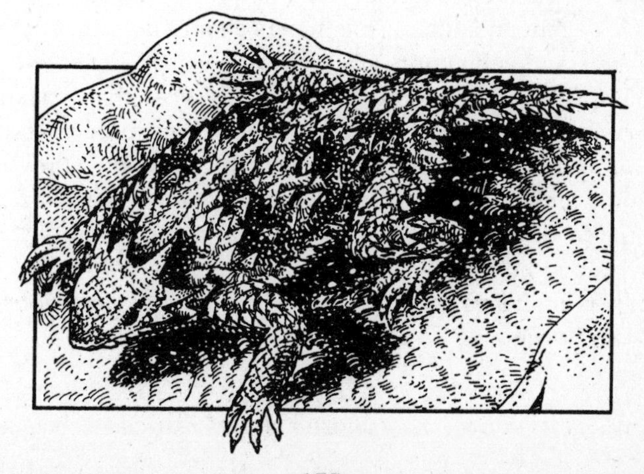

ture is threatened, it puffs itself up; this, conveniently, raises its blood pressure and thereby ruptures the capillaries near the corners of its eyes. Out squirts the blood—and it can hit its target up to seven feet away.

Is It True That Bees and Wasps Die after Stinging?

After they sting, bees die and wasps do not. In both cases, only the female stings, with the stinger a modified egg-laying tube, or ovipositor. The bees' stinger is a hollow sheath, with two barbed darts attached to a venom sac and poison glands behind. After a bee has stung in defense of the hive, attempts to disengage actually pull its abdomen apart, resulting in death within two days. The wasp's unbarbed stinger, used for offense—to subdue, paralyze, or kill prey—can be withdrawn and used again and again. (Picnickers should note that both yellowjackets and velvet ants are species of wasps.) These venoms are powerful and kill more people every year than snakebites do, usually because of severe allergic reactions to the histamine, serotonin, phospholipase, and melittin in them. Preventive measures include avoiding bright colors, wool or leather fabrics, perfume (even in shampoos), and sugary drinks; wearing shoes; and handling no food.

Other stinging pests to watch out for include scorpions, fire ants, and hornets. It's a tough world.

The Butterfly's Blooming

Watching a butterfly reborn from a caterpillar, the Greeks named it psyche, their word for soul. This exquisite transformation from a creepy crawly to a bright flyer has as its purpose a search for a mate.

The lives of some insects are divided into three stages: larva, pupa, and adult. A butterfly's chrysalis and a moth's cocoon, as well as the similar structures in other insects, are safe enclosures for the larva (a caterpillar, in the case of butterflies and moths), in which the larva can become a pupa and the pupa an adult. The sacs are woven of silk, with bits of hair, leaves, dirt, and chalky excretions for reinforcement. Within them the original caterpillar's body changes, and the butterfly's cells come of age. Nothing new is added, and almost everything is rearranged in this simulation of the miracle of time travel.

In trying to be born, a butterfly first ruptures its chrysalis. Gradually, its blurred wing pattern focuses, and its soft, folded wings spread and harden. This is the most elaborate visual display in the insect world and one well suited to attracting a mate, especially since butterflies have excellent ultraviolet vision and see even more patterns in each other's wings than we can. The color is created by tiny overlapping scales, some pigmented and some splitting the sunlight into its colors, reflecting back only some of them.

Decorated and able to fly, the former caterpillar sips some nectar and begins its fluttering hunt for a mate.

The First Flowers

A world awash in blue and green under white clouds might still be beautiful, but not nearly so much so as when adorned with flowers. Early grace notes in the age of the dinosaurs, the first flowers bloomed some 120 million years ago, truly a masterpiece of opportunistic evolution. They have come to attract attention and thus gain transport for their pollen and then their seeds courtesy of other living creatures, instead of relying on only the howling, blind wind. As the poet John Clare wrote, "Very old are the woods;/And the buds that break/Out of the briar's boughs/When March winds wake,/So old with their beauty are—/Oh, no man knows/Through what wild centuries/Roves back the rose."

Flowering plants, called angiosperms, are, then, a relatively late development on Earth. Long after the first green algae colonized the land more than 3 billion years ago, long after the shores became home to the first vascular plants that could move water and nutrients throughout their bodies, after the early ferns and the primitive forest trees, and after even the conifers, they came. Flowers entered a landscape already comfortable for winged and crawling insects, reptiles, primitive mammals, and the first birds. They succeeded evolutionally in cooperating with these first inhabitants and now are some 300,000 species strong, and the most highly evolved of the plants, because they became even more efficient at reproduction. Instead of scattering volumes of pollen or millions of spores to the wind at random, as many plants do, or trusting to the swimming of the sperm cells over the plant's surface to the egg cells, flowering plants adapted along with specific pollinators (such as bees) for pollen delivery among members of their species, and they came, also, to enclose the fertilized result in a relatively safe seed. Many also evolved a short life cycle, which allowed a speeding-up of their development. Some, notably the grasses, have since even been able to "reinvent" wind pollination for their own benefit.

Within a few million years of their development, the flowering plants had decorated the planet with roses and oak trees, forget-me-nots and corn, lilacs and rice, daffodils and onions, pollen-bearing flowers all. Their diversity of color and shape often gave them their names, phlox coming from the Greek word for flame, gladioli from *gladius,* the short sword of the Romans, tulips from the Turkish word for turban, *tulbend.* Their beauty caught the imagination of the Ancients, and they have not since ceased to capture sunlight with style.

This style has developed in full tandem not only with the insects but also with other pollinators. Though all mammals except the primates are either partly or entirely color-blind, insects have excellent vision. Not only are they attracted to color, but they carefully use it to distinguish the species of flowering plants for which they have developed a preference. Bees, which can see in the ultraviolet range of the spectrum beyond that visible to the human eye, use for pollen guides not only the colored centers of flowers but streaks and spots invisible to us. The tortoiseshell butterfly, for example, flies preferentially to yellow and blue flowers, while the large white butterfly avoids yellow and orange to fly to the blue and violet ones. White flowers attract mainly the night-flying insects and bats, too, as pollinators. Birds, good pollinators also, especially in the tropics, seem drawn mostly to red blooms. The colors of flowers have probably developed differently for this reason.

Fragrance, too, attracts some pollinators, notably, one kind of bee to tropical American orchids. So does heat, especially from flowers like the skunk cabbage and voodoo lily.

But vision, fragrance, and heat are not even quite enough to make flowers and insects a perfect match. Also needed is synchronized blooming, the blooming of various flowers at different times of the day and the year, to provide a full menu for their necessary bees. In springtime, trillium blows white in the cool woods. Hollyhocks are summer-hot against the garden wall. Chrysanthemums hold the last drop of yellow sunshine against a blue October sky. Poinsettias need only be urged a little to flash a merry red as late as Christmastime. And others, like roses and snapdragons, lie ever in bloom over the summer. The length of daylight, or photoperiod, has molded the lives of these and all flowers in the temperate zone, where daylight varies a great deal. (Tropical plants tend to flower for much longer periods but do follow the rhythm of moisture and dryness in their habitat.) The result of this coordination is three groups: those that bloom in summer when daylight hours are long (called long-day plants), those that bloom in spring or fall when the sun's ride across the sky is shorter (called short-day plants), and those whose flowering is largely unaffected by the length of the day (called day-neutral plants). The flowers have, essentially, divided up the world so that the bees can come to all of them.

A beautiful by-product of this synchronization is its appeal to people. Spring crowds our woods, yard crannies, bogs, and meadows with wild blooms of the violet, marsh marigold, and lady's slipper, all of which respond to less than twelve hours of sunlight. Summer brings to us the flowering of gladioli, clover, delphinium, iris, and garden plants like the beet and spinach, which require close to fifteen hours of light. Fall's shorter days usher in the asters, goldenrod, dahlias, chrysanthemums, and then poinsettias, which all need less light. And plants neither hurried nor delayed by the day's length include the carnation, dandelion, snapdragon, tomato, and rose, many of which may well have evolved near the equator, where they need not have matched their flowering to the angle of the sun. Response to photoperiod definitely sets the geographic range of a plant, after all; flowering in winter would make for a beeless, useless death.

Our own attraction to flowers much precedes written history. Evidence of once brightly colored bachelor's buttons and hollyhocks seems to have been found in a Neanderthal grave 40,000 to 60,000 years old. An ancient hunter with a crushed skull had been laid among them.

Aggressive Ducks

These tough birds don't just dabble peacefully for wild celery in bucolic farm ponds or sylvan pools. Canvasbacks, the fiercest fresh-water ducks of all (at least in South Carolina, port of their winter territory), shove and lash and bite one another to establish and then defend their small inland feeding areas from the Atlantic and Gulf Coasts to Mexico. One of many North American fresh-water ducks that dive for their food instead of dabble, the canvasback guards a rough circle of 20 to 30 feet. The sound of the male is a low croaky growl, and this duck threatens first; then, if the other duck doesn't back down, it attacks.

Can Some Animals Really Grow New Body Parts?

Lizards can sprout new tails, frogs and toads grow new legs, and starfish and octopi generate new arms. Rabbits and cats and some bats can repair holes in their ears, and some earthworms grow whole new body halves (including a new brain for the new half that lacks one). Also, mice, monkeys, children, and even some adults can grow new fingertips. Cyclically—which seems less incredible—deer can sprout new antlers, and elephants, sharks, dinosaurs, as well as other reptiles, endless new teeth.

These are only a few examples of a vast, natural fix-it system that is only barely understood, but clearly goes beyond small wound repair. Simpler creatures seem more adept at it. Among the more complex creatures, amphibians seem especially good at regeneration. In the case of mammals, the capacity is more limited and may even have a different cause. Regeneration depends upon the place of the amputation, for example—and regrowth is most easily accomplished in younger animals. Further back, in the embryonic stage, flexible new growth is possible, even among human beings—a very early cellular split that works well, after all, is identical twins.

Among children, and even more rarely among adults, there are documented cases of the self-repair of fingertips. This marvel of regeneration in our species was unknown until fairly recently, because what had been standard surgical treatment—placing a flap of skin over the wound—seemed to interfere with the regrowth. The wound should only be cleaned, perhaps stimulated electrically, then left to grow again, scientists now say. A new fingertip may come.

Feats of Bird Migration

A bird of the iced sunlight, the Arctic tern migrates nearly 25,000 miles a year from the northern land of the midnight sun to Antarctica's long summer, and back again. It is said to be the bird that hardly sees darkness, the longest direct distance migrator. The short-tailed shearwater samples nearly 22,000 miles of Pacific shore in its loping yearly flight. Even the tiny ruby-throated hummingbird manages a more than 1,000-mile migration, 500 of them over the open waters of the Gulf of Mexico, its travel from warm to warm again. Migratory birds use their own combinations of the stars, the winds, landmarks, an inner sun compass, and the Earth's magnetic field to find their destinations, their journeys all triggered by the changing hours of daylight in the turning seasons.

Migration among tiny songbirds, and certainly among shore birds, is common. About two-thirds of the songbird species that decorate the northern United States engage in summer travel between 250 and 2,000 miles each way in their annual voyages; some species add 50 percent to their body weight before doing so. These trips are so relentlessly efficient that if a blackpoll warbler burned gasoline instead of its own body fat, it would be getting 720,000 miles to the gallon. Shorebirds are often even longer migrators. Even the smallish white-rumped sandpiper travels 10,200 miles a year in migration. Traveling these vast distances perhaps evolved gradually, as the last receding glacier opened up new and richer summer feeding territories farther north—and birds found them worth the trip. Or, since the continents have drifted across the face of the Earth, the birds may be migrating farther now than they once had to.

The stars are the guide of night migrants from robins to mallard ducks, blackpoll warblers to European warblers and indigo buntings. The bunting, for example, demonstrably knows the patterns of several constellations, including the Big Dipper, Little Dipper, Cassiopeia, Cepheus, and Draco, all of them close to the North Star and so circling tightly a point near it every night. (Constellations farther from the North Star move in wider circles and are, therefore, less reliable. Some dip below the horizon nightly in their circles and even disappear for months on end as the Earth faces different areas of the sky throughout the year.) As a bunting matures, it learns first to find the north-south axis of star rotation, now very near the North Star, and next learns to recognize several constellations around it. On cloudy nights, some night migrant species do tend to get lost, while other skysailers fall back on different directional cues.

For both day and night migrants, the winds provide help, too. Along the Atlantic coast in the fall, for example, some 100 million songbirds and small shorebirds wait to begin their migrations. Some, sensitive to changes in barometric pressure, watch for a cool high-pressure system coming in from the west; these bring north/northwest winds that can carry them southeast over the Atlantic toward Bermuda, then to the Caribbean and South America, as they pick up winds moving southwest. Many hug the coastline to Florida, then turn southeast to follow the Caribbean Islands. The first Atlantic wind is also usually a reliable sign to the birds that no tropical storms or hurricanes are approaching the North American coast. It is a high altitude wind, too, forcing the birds to fly at altitudes of up to 21,000 feet, much higher than they would if the wind held no advantage for them. Other birds use winds to detect land and sea odors and orient themselves accordingly. And still others use the wind along with a sensitivity to tiny changes in barometric pressure.

Landmarks are used by birds, too, basically for general direction and then again for last-minute orientation very near their destinations. Broadly, most North American birds use one of four vast "flyways," each complete with its own consistent landmarks: the Atlantic Coast flyway, the Rocky Mountains flyway, the Pacific flyway, and, the most traveled, the Mississippi River flyway. Once very close to their destinations, usually an area of only a few square miles, the birds seem able to recognize a mountain, a stream, or a grove of trees.

Of all the directional cues used by day migrants, the inner sun compass is perhaps the most important. It works even when winds blow flocks off course and when landmarks change. It is for day migrants what the stars are for migrants of the night. Birds using it range from petrels and albatrosses to warblers and starlings. Before leaving, they will wait out fogs and storms so that they may take their initial measure of the sun. Though its angle changes about 15 degrees every hour as it traverses the sky, the birds compensate for this, seeming to rely both on the sun's apparent horizontal movement and its altitude, comparing them both to remembered angles at their two seasonal homes. They require only a partial blue sky, not a full sun, since they can see polarized light through clouds. Exactly how a bird can migrate thus by the sun is still a mystery.

Just as difficult to fathom is the magnetic sense that birds are now known to have, at least once they have had some experience with migrating. Robins, for example, fall back on this sense only if they cannot discern stars or sun. Homing pigeons seem to use it only for short-term navigation, along with their sun compass and their sensitivity to very low intensity sound waves. Some fish, insects, and lower organisms are also known to have this sensitivity. One scientist has even tested people and has found some preliminary evidence that we may have it.

Birds, in general, seem to have more than one directional talent for migration, as befits a task that absolutely must be completed under a wide variety of weather conditions. Their readiness for it is also seen in the build-up of body fat and in nocturnal restlessness as they sense the changing length of the sunlight hours. And the latest evidence suggests that they are somehow endowed with an internal, genetically controlled, awareness of a yearly cycle.

There are variations on the migratory theme, too. The first is vertical migration, accomplished by birds that migrate up and down mountains. Some 1,000 feet lower on the mountain, it is instant summer again, the equivalent in plant life of about 600 miles south. Some birds are also nomadic—taking regular short flights to follow food—or vagrant, traveling both shorter and long distances, depending upon the food supply. Only sedentary birds, like the house sparrow, golden eagle, and nuthatch, remain in one place all year long. In some species, such as the blue jay, the younger birds migrate, while the older ones are sedentary. In still others, only the birds in the northern part of the species' range migrate. And a few species, like the penguin, migrate by swimming, an arrow bound through a different medium.

Insects That Are What They Eat

You are what you eat, they say, and this is literally true of the caterpillar that develops later into the geometrid moth. Those born in spring and summer initially look alike and, in experiments, are even found to have been brothers and sisters. But the springtime ones feed on oak catkins and quickly begin to look just like that part of the oak, while summer's children eat leaves and come to look like twigs. (Caterpillars all do nothing but eat, sucking food up through the proboscis.) The result in both cases is camouflage. It seems to be a difference in tannin levels in the two diets that creates this dramatic difference over a transparent, colorless template, which extends to the head and jaw shape of the caterpillar, as well as its color.

This "developmental polymorphism" is also characteristic of some other caterpillars, pupae, butterflies, plant hoppers, aphids, and water striders, but not to this startling degree.

Sun Worshippers

Following a golden sun across a blue sky are the sunflower's flowers, the mallow's leaves, and an entire alpine flower. The mallow, a common weed, follows longer, since the sunflower stops once its flowers have matured. Then it begins to face east all day long.

The sunflower's early path is more dramatic and lovelier than the mallow's and is, in fact, mythic. In a Greek myth, a maiden in love with the Sun god, who did not return her love, sat outside to watch him travel across the sky each day. Finally, she herself was changed into a sunflower.

To a plant, the sun is indeed a god—a force without which life would be impossible. Many kinds of plant movement, called tropisms, are responses to changes in its light or heat. The movement of sunflower and mallow is a type of tropism called phototropism. Botanists are not yet sure whether it is caused by changes in turgor pressure in the plant's cells as they swell and then lose water in response to the sun's heating opposite sides of the plant unevenly, or by the pooling of a plant chemical called an auxin on one side of the plant body, elongating cells and thereby causing the plant to bend. Both are common mechanisms behind various plant movements.

Other kinds of plant movement include nutation, the family of circular movements that includes the twining of vines as they grow; various nastic movements (all independent of the direction of stimulus), such as the opening and closing of the wood sorrel in response to the intensity of sunlight, or the curling up and drooping of the mimosa plant in response to touch; the "sleeping" movements of flowers like the dandelion, morning glory, tulip, and crocus; the various tropisms (direction-dependent movements), such as geotropism, in which plants respond to gravity, thigmotropism, in which they respond to contact with a solid object, and hydrotropism, in which they bend in response to a water source; and the most common kind, phototropism, in which a plant such as the geranium or marigold leans toward the light. The sunflower's and mallow's daily moving paths put them in a distinct plant minority.

Sunflowers have long been considered special, much more so than the mallow. Priestesses of the sun in the Inca religion wore golden sunflowers on their breasts. American Indians used its flowers as a dye for pottery, basketry, and ceremonial body decoration, its roots as medicine, and even its stalks to make flutes. Throughout time its seeds and oil have been considered good food, spun from the sun. Curiously, no one has had much to say about the mallow, a common weed of roadsides and pathways, nor of the relatively insignificant alpine plant.

A Reptilian Divining Rod

A person in search of water in the lowlands of Sri Lanka, Indonesia, the Philippines, and throughout Southeast Asia, in general, can follow a water monitor. This huge black and white carnivorous lizard—up to 8 feet long and 50 to 75 pounds heavy—is, unless it is heading for a termite mound in which to lay its eggs, almost always going to get a drink. It is never far from one. (And it doesn't want to drink you.)

A Jail Bird

It is not spouse abuse but a need for protection of the eggs that has led the great hornbill male to feature an unusual architectural behavior. He finds, and then excavates further, a large tree hollow (large enough to fit a bird almost a foot and a half long), then seals most of it off with bits of mud brought from a nearby water hole. After he mates with the female, she manages to squeeze in, and the male seals off the rest of the hollow, leaving only a breathing hole through which he

will feed her. She stays in there for a full 40 days. Once the eggs hatch and the young begin to grow, the female leaves, closing off part of the trunk hollow behind her. The little ones seal up the rest of the hole themselves, then are fed through the opening that remains until they are large enough to have their freedom too.

Astounding Hibernations

In 1768 Samuel Johnson told Boswell, his biographer and friend, "Swallows certainly sleep all winter. A number of them conglobulate together, by flying round and round, and then all in a heap throw themselves under the water, and lie in the bed of the river." This was, then, an accepted view. Actually, there is only one bird that hibernates—Nuttall's poorwill, a western North American relative of the whippoorwill.

But, all over the world, come the cold season, living things are entering their long sleep. A snake slithers into its burrow for winter, some frogs bury themselves in the pond bottom. Bees hide in their hives, warm amid honey and a fluttering of wings. The mourning cloak butterfly flies into a woodpile. The poorwill slips into its cleft in the desert rocks. A fat woodchuck's pulse quiets almost to silence. Bats fly into caves to cling to the ceilings.

Hibernation—a deathlike sleep—is quite a drastic solution to the problem of finding food in the winter—and it is not a common one; migration is a much more typical animal strategy, and toughing it out is more common still. In fact, hibernation is a phenomenon mostly of the midlatitudes (higher north most animals would freeze to death underground or at least run out of body fat).

Most common among reptiles, amphibians, insects, and anything that eats insects, hibernation must have been long in coming—after all, a half-winter hibernator may be a dead hibernator—but whole-winter hibernation would have been well rewarded once it evolved. The frog, quiet in its muddy pond bottom, and the snail in its shell sealed with slime use very little energy at a time when their stores cannot be replenished. Insects, whose hibernation is called diapause, may actually use no energy at all, freezing solid like the carpenter ant, its tissues saturated with glycerol, a type of natural antifreeze; in this way the carabid beetle can survive an Alaskan winter where temperatures may drop to −40 degrees Fahrenheit. Many insect species overwinter only in egg or larva form. Honeybees, originally from the tropics, gather in tight circles in their northern tree-hole hives, generating heat by vibrating their wing muscles and sipping honey. The mourning cloak butterfly lives through the Minnesota winter buried in weeds, logs, or a brush pile, all insulated by snow. The poorwill is even called the sleeping one by the Hopi Indians because of its hibernation.

Among mammals, the physiology of hibernation is well developed and eerily efficient. The animal first eats furiously—a woodchuck may consume a third of its weight in plants in a single autumn meal. Some hibernating bats mate before hibernation, then store the sperm for later pregnancy. The animal then slips gradually into a suspended state of animation. The pulse slows to a few beats per minute, blood thickens, kidneys nearly stop, and the deep body temperature drops nearly to freezing. A jumping mouse in this state, curled up in a ball in its meadow, can be brought inside and rolled across the floor without awakening it. This state lasts long, too—about five months for woodchucks, seven months for a ground squirrel in Minnesota—though many small mammals awake periodically to eat from their food store.

Some scientists suggest that the first mammals to hibernate were those that originally lived in very hot climates; they spread north gradually, adapting to winter-sleep as they moved. Others say that this strategy in mammals is a vestige of their early reptilian and amphibian ancestors, whose body temperatures typically and easily took on that of the air or ground or water around them. This much is certain: Without hibernation, the northern world would be home to fewer species.

Animals that fall short of hibernation—that rest for shorter periods and whose body temperatures could not safely sink so low as those of the hibernators—are considered to be in a state of torpor, or "carnivore lethargy," instead of true hibernation. Bears, squirrels, badgers, raccoons, and even some chipmunks in most climates fall into this category (in colder climates, chipmunks hibernate). The animals take on enough fat to be able to skip a few days' food, then sally forth outside again every time temperatures warm slightly. Bears even wake to bear their young in winter, remaining with them in the den. Some birds, notably swifts, goatsuckers, and hummingbirds, enter torpor even during relatively short cold snaps. Huddled together, they can die if temperatures do not rise soon. Hummingbirds may even enter this state on an ordinary night if their inner energy reserves are low.

Oddly comparable to hibernation is aestivation, an animal strategy for avoiding dangerously hot, dry periods in tropical climates. At the beginning of the dry season, the African lungfish burrows into the swamp bottom, cocooning itself in a moist layer of slime. So it stays, sharing a strategy with many of its distant northern cousins. This dry state of suspended animation, called anhydrobiosis, is possible because the creatures produce increased amounts of zinc and a sugar called trehalose as they are drying; it somehow insulates body cells and thereby prevents damage from dehydration. In their dry state, metabolism seems to stop completely.

Other, less drastic measures work well in the desert, too. One flower, the "living stone," hides most of itself underground; the spadefoot toad lives entirely underground for eight to nine months of the year, and the tiny kangaroo rat comes out only at night, hiding in a plugged burrow all day, excreting almost no water. Also coming out only at night are lizards, snakes, jackals, and foxes. The cacti of the desert use their spines to help screen the sun away, to reduce evaporation by breaking up the hot wind, and to collect rain and dew, allowing it to drop gently on the plant. Even the names of many of them sound spine-prickly: pincushion, fishhook, eagle claw, hedgehog, and porcupine. Some cacti can wait years for rain, then sprout, flower, and bear seeds in as little as two weeks. One desert quail, too, waits for the rain and the vegetation it creates before it will mate. And spring will come.

Creatures That Never Sleep

Bullfrogs don't seem to sleep at all, a good many fish never slumber, and the "resting time" of all amphibians and reptiles does not seem to resemble our sleep at all, or even that of other mammals and birds, which is quite similar to ours. Predators tend to sleep longer than prey, and animals with faster metabolisms usually, but not always, sleep longer than those with slower metabolisms. This information has been established with some difficulty, by wiring the creatures to detect their brain waves; so it is not surprising that no one has been able to accomplish this yet with a fish or an invertebrate.

It is known, though, that many fish would asphyxiate if they did not swim constantly, moving water past their gills to get oxygen into their bodies. The mackerel is an example of this human nightmare of ceaseless swimming. But even though they have no eyelids, some fish do indeed seem to sleep: The parrot fish secretes a mucous "blanket" to surround itself and at least rests, as do yellow perch and mullet, on the bottom; some wrasses bury themselves to look like rocks; certain sharks sleep in underwater caves.

Some typical sleep times: dolphins, with one eye open for about 1 hour; sharks, about the same; the small roe deer, about 2 1/2 hours; the horse, which likes lights at night, less than 3 hours (it is one descendant of all the wild, grazing animals that probably had to run regularly from predators); elephant seal pups, 8 to 10 hours (some of which is spent without breathing at all); chimps (which have the same sleep stages as we do and even seem to dream in almost the same cycles), about 8 hours; predatory birds, 11 to 12 hours; and the opossum and bat, about 20 hours every night.

Within this surcease of consciousness, which, as Shakespeare says, "knits up the ravel'd sleave of care," all we humans dream. And so, as best as can be figured out, do all mammals and all birds except—of all things—the spiny anteater bird of Australia. It, of all birds, has no dreaming brain waves in its sleep.

Dinosaur Descendants

The entire kingdom of birds is descended from the dinosaurs; feathers from their scales and wings from the second fingers of their claws. The first bird probably flew more than 150 million years ago, before flowers had evolved. By 120 million years ago, they graced the skies all over the world.

The first one known was an ungainly small animal called archaeopteryx. It laid eggs and had feathers, though, and so it was a bird. Crow-sized, it wobbled over bushes and barely maneuvered among trees. But its asymmetrical wing feathers show that it could fly. (Birds that don't fly have symmetrical "flight" feathers, instead). In the days when nothing else existed that could catch flying insects, its awkward flight was good enough. In between airborne forays, it probably used its strong claws for perching and ran along the ground capturing small animals and low-flying insects. The trees were probably safer from the predatory dinosaurs, and, later, from the mammals, encouraging further stages of its evolution. Archaeopteryx surely did not sing; instead it may have croaked, grunted, moaned, or hissed. And like the reptiles it came from, it had small teeth and a tail. Not lovely, it was utterly unlike the bluebird, who, as Thoreau says, "carries the sky on his back."

The barely flying archaeopteryx may have evolved from a creature that glided down from the trees or one that jumped up from the ground, a matter still in controversy. In both cases the transition to a wing shaped for flight seems difficult. The fossil record is silent here. But at least one thing is known: A jump can increase one's foraging space considerably, and a very small bit of lift can make jumping and landing from a jump much less awkward. So intermediate steps that could have encouraged the evolution of flight from the ground up at least look possible.

This odd bird was not the first creature to evolve flight. Flying insects had already colonized the Earth, and the reptile family called the pterosaurs had found the advantage of flight after that, appearing even before the true dinosaurs. These creatures ranged from the evolving pteranodon, with a wingspan sometimes as broad as 40 feet and perhaps the largest living thing ever to fly, to the smaller pterodactyls. They were clearly reptiles, though, leathery wings and all, not birds.

By the time the pterosaurs, dinosaurs, and archeopteryx had all become extinct, they had left their legacy. Though the stages of their descendants may be hard to trace—feathers are rarely fossilized—the latest ones are flying now everywhere in their colorful, songful profusion.

Can Black Widow Spiders and Tarantulas Actually Kill You?

The venom of a black widow is more virulent than a rattlesnake's. The sheer breadth of a dark, furry tarantula can be near ten inches. Yet neither of these spiders presents an overwhelming danger to people; it is smaller prey, such as birds, lizards, and other insects, that the venom is designed to kill.

The bite of a black widow spider is not pleasant, however. Its neurotoxic venom can cause nervous system disorders including temporary paralysis, muscle spasms, abdominal pain, and some blood and tissue cell breakdown. In very susceptible people, it can occasionally cause death. Before injecting its venom through sharp fangs, the black widow wraps its victims in silk (well, the smaller ones), then adds more silk while its toxin dissolves the prey's body from within.

Tarantula venom is much less harmful to people, and even when handled, this spider rarely tries to bite. The venom is neurotoxic like the black widow's, though, and can cause paralysis and spasms in heart, lung, and other muscle tissue. North American tarantulas have a leg span of about five inches; only the largest species, in the Amazon, approaches ten inches across and can weigh up to three ounces.

In addition to these, almost all species of spider are venomous to some extent. Even our "house and garden" variety of spiders use venom on the small insects caught in their webs, a poison that usually includes a preservative to keep their dinners from rotting. The funnel-web spiders of the tropics are the most poisonous of all, injecting toxin from two immense fangs.

Besides spiders and snakes, venom is an even more common

weapon of creatures of the sea, used by some sharks and many sponges, sea anemones, jellyfish, scorpionfish, and ratfish. On land, ticks, wasps, ants, bees, and scorpions also pack powerful venom. Whip scorpions even squirt it from their tails to kill prey from cricket-size up to lizard-size. Fire ant venom can destroy an uncomfortable number of blood vessels in small mammals. And the Australian bull ant's venom contains histamine (like that of bees and wasps), which can cause shock, blood pressure loss, and death, even in people (if they are susceptible).

But only the famous tarantula actually has a dance named for it. The tarantella, a frenzied dance, is said to be the only way to recover from a tarantula's bite. It is danced ritually still, in village squares all over Italy, a legend that will not die.

The Largest Hopper

The hopping flea is famous but small, and the giant kangaroo, ten feet tall, is extinct. This leaves the ordinary red kangaroo, hopping its way across Australia, as not only the largest living contestant but the prizewinner for the longest hop. A record hop from one of them was 42 feet long, and even an average hop is 25 feet or more, with the animal able to accelerate up to 40 miles per hour for brief distances in the process. They can also jump as much as 10 feet high, to clear a woodpile, for example. Large males can weigh more than 200 pounds, and they sometimes fight each other with their front nonhopping paws.

The hop of the kangaroo is not only record-breaking but is an amazingly efficient way to move, at all but the lowest speeds.

Like a child on a pogo stick, the animal stores "elastic" energy, then uses it to suddenly spring forward. Instead of the mechanical spring of the pogo stick or the chemical springiness of a rubber band, though, the kangaroo uses the tendons of its hind legs for its intermittent and excellent energy storage.

Speed is crucial for this efficiency. Below about three and a half miles per hour, kangaroo hopping is distinctly inefficient, and at these speeds adult kangaroos lumber along awkwardly, using their tails as extra legs to keep their balance. At medium-hopping speeds, from about 9 to 25 miles per hour, the kangaroo becomes very efficient—it uses less muscular energy to move its body weight than another animal with two legs (or even four) would. In fact, the kangaroo takes about the same number of hops per minute throughout this speed range, just using longer and longer strides in rhythm. Within this range, at about 16 miles per hour, kangaroos are at their ultimate efficiency, and an adult animal can hop for miles. At higher speeds—25 to 30 miles per hour or more—the animal must take longer and more frequent hops and tires more easily.

If hopping is so efficient, why has it not evolved in many other animals? It is indeed associated in our imagination mostly with the hopping marsupials, which number about 45 species, including five or six species of large kangaroos, smaller wallabies, and the smallest, rabbit-sized, kangaroos, all having evolved this skill in Australia. But it is also fairly common among small desert rodents around the globe, such as the jerboa. Hopping seems to be quite ideal for smaller animals because they can switch easily to a four-footed gait for lower speeds. (And it works for the flea in a different way—a creature that can jump as high as 130 times its own height.) The large kangaroo cannot switch to four feet, because its hind legs are too large and its forelegs too small, to walk or run. It is, however, the only hopper that weighs in at more than ten or eleven pounds, and it looks very impressive in action.

Meat-Eating Plants

These plants eat spiders, butterflies, mosquito larvae, beetles, tadpoles, flies, ants, and even, very occasionally, a small mouse. The nearly 500 species of carnivorous plants of the world have evolved into their odd menu—of mostly crawling and flying insects—because it has allowed them to live in nutrient-poor places like heaths, bogs, rock crannies, inhospitable forest soils, small pools of water, and even marl (a very crumbly soil). They are equipped to conduct photosynthesis, too, and can rely on it increasingly if the soil improves, but they never grow big and strong unless they have "meat."

Charles Darwin called some of their leaf systems "temporary stomachs," since their death-dealing leaves, which usually lure the prey and trap it, include special glands that secrete digestive juices and feature absorption glands to move the food's nutrients throughout the plant. The traps come in three basic designs, akin either to a slippery-lipped pitcher, sticky flypaper, or a steel trap.

The pitcher plants are of the first type, a group called "pitfall catchers." Sparkling with color on the outside and sweet with nectar on the inside, the leaves can easily seduce an insect to its lip. There the smooth, steep, waxy, pitcher-shaped wall allows the hapless creature no good foothold, and stiff, bent hairs keep it from even the prayer of crawling back out. The insect slips to the bottom of the pitcher; there it drowns in rain water and digestive fluid and is gradually consumed by the plant. An occasional small tree frog and even, rarely, a mouse have been found to have met this fate in the larger pitcher plants, which can easily hold a pint of fluid. The largest of all, found in Borneo, holds three and a half quarts of water. (That one features spiders who wait inside to capture the smallest prey, and provides water for monkeys to sip.) The fluid of pitcher plants is fresh and potable, except for the digestive juices and insect parts that settle at the bottom. Quaffers should also watch out for the pitcher-plant mosquito that lives inside (escaping death by not alighting), the aforementioned pitcher-plant spider, and the larva of one moth that uses the pitcher as an individual hibernating place.

Next is the type of meat eater that uses a sticky surface to perform its tricks, a group called the "lime-twig catchers." One of them, the European flycatcher, actually is used for home flypaper in Portugal. Another, the sundew, is a halo of tiny hairs glistening like dewdrops, luring insects with its color and fragrance. When an unlucky one lands, each hair secretes a bit of sticky slime to trap the prey. As the insect struggles, it only tangles itself in more of the hairs, which curve around it like threads of adhesive. These hairs, actually glands, then secrete enzymes for a few days to digest the food, reabsorbing them when the meal is over. One naturalist, walking a vast two-acre meadow of sundews in Britain, saw a special carnage. Some 6 million migrating butterflies, all cabbage whites, had tried to come to rest on the plants. And they were virtually all being eaten, each sundew consuming several of the delicate butterflies.

The steel-trap style of carnivorous plants is perhaps the best known, mainly because it includes the Venus flytrap. The leaves of this plant, brandishing tiny "teeth" and touch-sensitive bristles, can snap shut from the hinge around a hopping or crawling insect or spider. When the prey first touches the tiny trigger hairs, the plant's cells nearby begin pumping themselves up with delicate yet energetic electrical pulses. In what may be the fastest cell expansion in the plant world, the final snap shut is accomplished within one to three seconds. Inside, the insect is locked in a reddish, barred cage of trigger hairs. There it somehow chemically stimulates the flow of the plant's digestive enzymes. The meal is completed—and the leaves reopen for business—in anywhere from ten hours to several days. By that time, only the legs, wings, and shell of the victim remain, all light enough to blow away in a breeze. A relative, also operating a "steel trap," is the bladderwort. Mostly water plants, these species have small elastic-walled "bladders" that lie like two hinged suction cups just below the water's surface. When a water flea or mosquito larva triggers hairs at the narrow opening, the mouth opens, allowing the prey to be swept in with a current of water. Capture happens faster than the eye can see, in about 1/35 of a second. Then the leisurely meal begins, several days of dissolving the insect for all to see through the translucent green bladder's walls.

It's all in a summer's day, nature green in tooth and claw.

The Shy Octopus

This creature can bite—even with some poison in its saliva—and it can pull your diving mask off and grab at your neck. But all this is only because it occasionally seems to confuse divers with intruders of its own species. Once set straight, the shy octopus will gladly let you go. And if it doesn't, you can always stick a thumb into the opening under its head and turn it almost inside out. Not fun but not hard, either, and it works.

So the image of a gasping, choking swimmer being pulled toward the coral reef lair of a giant octopus, never to return, is a vast exaggeration. Daunting the creature is, but deadly it is not. The very largest octopus ever found did indeed have an arm span of 32 feet, but the common one stretches only to a foot or two and has a very small body. Even when the typical octopus is agitated, it usually just squirts black ink or a blast of plain water, or it turns colors from grayish to pinkish to green to reddish, and finally to an iridescent sheen. These are not the stuff that nightmares of the sea are made of.

An octopus is likely to distinguish a flailing diver from a fellow octopus fairly quickly, too, because octopus intelligence is quite high, probably the highest of any nonvertebrate. They have been known to open jars to get food, go through mazes, distinguish among objects of different colors, sizes, and shapes, even move from tank to tank upon human command and then remember the experience.

They are interesting creatures otherwise too. With eight tentacles equipped with suctionlike cups, they can use them to grab food, propel themselves backward in spurts, fight, and mate. The tentacles, if severed, can even regenerate. And the creature can sometimes use them to crawl briefly out of the water to snatch a crab. Favorite foods are mussels, scallops, clams, lobsters, and crabs, usually caught by changing to a good camouflage color and ambushing the prey. They can also build simple shelters to hide themselves or to hide their 50,000 eggs; the female waits alongside, guarding them, for about six months without eating, then dies. An octopus has three small hearts. It is a mollusk evolved to have lost its shell, but not to have become excessively dangerous.

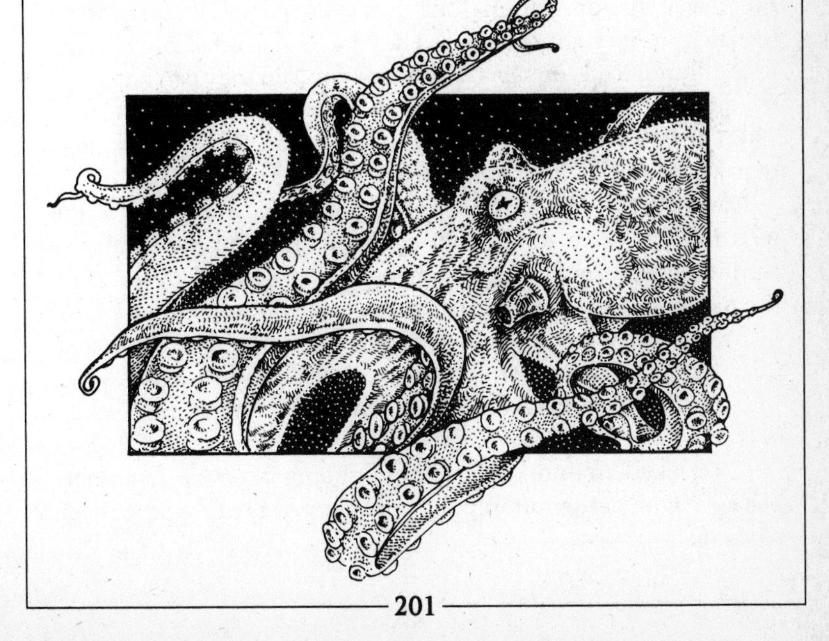

The Smartest Birds

Big and black and clever, ravens are the wilderness cousins of crows. Other members of their family are jays and magpies—and they are all smart.

The intelligence of ravens shows most splendidly in their flexibility. They can recognize a gun from afar and will set up a raucous cacophony, yet they can behave very "tamely" with unarmed people, snitching garbage with aplomb. If adopted by humans when young, they will not only become loyal pets but make quick associations; one flew to its owner, an ornithologist, with a clothespin, happened to receive from him a bit of special food being prepared, and rapidly departed for another clothespin. They can also be taught to speak at least as well as parrots. The famous ethologist Konrad Lorenz said that his pet raven was "the only animal that has ever spoken a human word to a man in its right context." They have, it seems, about sixty-four different calls or signals of their own, and they can be taught to count with comprehension, as well as to speak human words. They are also particularly playful, doing acrobatics at heights that would put a champion gymnast to shame.

The memory of a raven is prodigious. One was seen in the wild to hide more than 1,000 eggs for food and eventually to retrieve every single one (each had been marked by an ornithologist). Ravens will return to the same well-hidden nest for years, mated for life. One usually inseparable pair was seen quarreling, then separating for hours after one dropped a fish into the river by mistake. They teach their young hunting skills and the fears necessary for foraging near people. And a flock once attacked an entire apartment building in Israel, seemingly because of an earlier attempt by a resident to take a young raven as a pet.

The birds are opportunistic, robbing other birds of their prey, finding new human structures such as telephone poles for nests very quickly, and locating new human settlements in the Arctic as a source for food. Their most recently discovered "talent" is the tendency to recruit other ravens to come with them to a carcass for food; this highly unusual behavior seems to have evolved to help younger birds manage to get a share of food from within the territory of an older, mated pair—a whole group of them cannot be repulsed by the pair easily.

Up in the Arctic, where many live, they have not evolved to require any ruse, such as turning white for winter, relying on their wiliness for protection instead. Their range is, in fact, extending instead of shrinking, with ravens found from the high Arctic to Mexican deserts and from northern Europe to the Sahara. They corroborate the theory of many evolutionists that a varied diet and intelligence are associated.

These birds are large for pets, with wingspans of more than four feet. A crow might be more convenient and seems to be almost as smart: One was known to walk its owner's dog by picking up the leash in its beak. No bird brain.

How Well Can Birds See?

Birds specialize in vision the way that insects do in smell, fish do in touch sensitivity, and some mammals do in hearing. They see much better, at both shorter and longer distances, than people. In fact, if our eyes were proportionally as large as those of birds of prey—the raptors—they would weigh in at several pounds each. An ordinary swallow can zigzag and scissor after insects too tiny for us to see at all. An owl in dim light sees about ten times better than people do (and just as well in daylight). A harpy eagle can see well enough to fly through the dense South American forest at 40 to 50 miles per hour. And even a house sparrow has twice the visual cells in its central retina that people do.

Because birds, especially the raptors, can see so well, they excel at hunting. One bird or another eats from every group of living creatures—they consume invertebrates, fish, amphibians, reptiles, other birds, and even mammals; the latter include the young antelopes eaten by martial eagles and the small monkeys grabbed by crowned eagles, as well as the small rodents consumed by hawks and owls. Even a robin peering with a cocked head in the direction of a worm is not near-sighted but merely getting a better angle from eyes that are placed on both sides of its head.

Within a bird's eyes—which often weigh more than their brains—are crowded hundreds of thousands of rods and cones that work well in several different intensities of light. They also have a semitransparent extra eyelid, called a nictitating membrane, which can function as sunglasses, a wind visor, and aqua goggles, as well as helping to clean the well-equipped eye.

A Drink in the Desert

"Water . . . water . . ." gasps the scruffy prospector in the old movie. Beard haggard, clothes tattered, he drags himself over outcroppings of rocks, then lunges at a large cactus. Will it save him from death by thirst? Can you rely on a cactus when caught with an empty canteen on a desert hike?

Yes. Though most cactus fluid is a bit slimy, bitter, and hot, it is not poisonous, and it is wet. The best choice for thirst-quenching is the giant saguaro—the one that looks like a thick, leafless tree, arms up—or a saurian cactus. Depending upon their size, these cacti can hold water for up to several years. The more recent the last rainfall, the more full its accordion-pleated skin will look—and so the more fluid you will find when you cut into the plant.

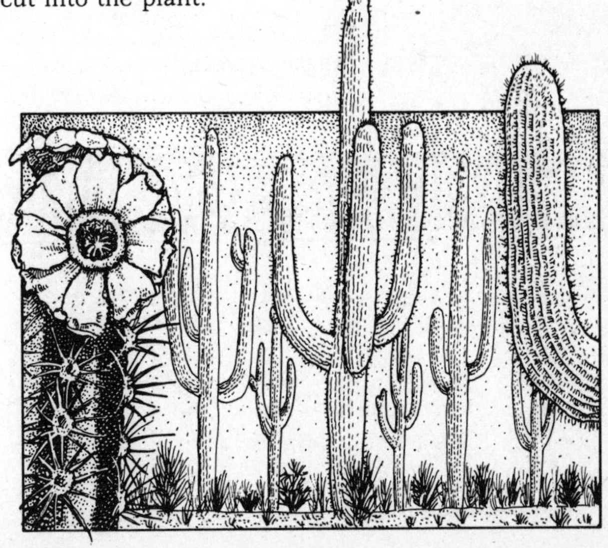

Cacti hold precious desert water very well. Their cells are like sponges: large, thin-walled, and widely spaced. Their roots are a veritable mesh, too, drawing practically every drop of water from the ground. And they have also evolved many ways of preventing the escape of life-giving desert moisture. The first is to have dispensed with leaves, which allow other plants to "sweat" and to breathe easily in a constant water exchange (a system that would be very bad for a cactus). They usually respire only at night, too, when opening their pores (called stomata) results in less water loss than it would during the heat of the day.

A traveler can find more water locked in other kinds of plants elsewhere, though. Lianas, the family of thick, clinging, twining vines that includes such orchids as the vanilla plant and philodendrons, are especially wet. Cutting into the stalk of a very thick one yields abundant fresh drinking water. So does the traveler's tree, a member of the banana family that holds fresh fallen rain water in its large fan-shaped leaf sheaths, and the bamboo within its segments. Unfortunately, these are not desert plants but rather rain forest inhabitants. And how many of us would be struggling on for days without water in the middle of a rain forest?

How Well Can a Dog Smell?

bloodhound is the main marvel, able to smell one drop of blood in five quarts of water and to detect the scent of fatty acids 1 million times better than a person can. Its owner's footprint still smells fresh to the dog 24 hours later. But even the average pooch has 1,000 to 10,000 times the nasal membrane surface that humans do and at least twenty times more olfactory receptors. The best of the smellers seem to be breeds with the longest or broadest muzzles.

Though they have lived with us as domestic companions for more than 10,000 years, dogs live in a world of scents that we could never enter and hardly imagine. They can distinguish many smells at once, separate them, select one to follow while ascertaining the direction of the source (by the relative strength of the scent in the different directions), associate smells instantly with past events, even those that occurred almost a lifetime ago, and learn the outdoor smells of a route while traveling in a car. One record-breaking dog, a fox terrier, found its way home over 1,600 miles of Australian territory, a journey still not explained. When a smell is faint, yet pleasurable, a dog will sometimes close its eyes to concentrate better. And even extraordinarily faint smells will wake it up if they signal either danger or pleasure. Especially strong smells like tobacco smoke can make dogs sneeze, but they do not get colds as we know them.

A dog's sense of hearing is also quite acute—it can detect frequencies almost twice as high as we can. And its vision is respectable. In daylight it sees about as well as a moderately near-sighted person and, on dark nights, better than people do. It is almost but not quite color-blind.

Back to smell for two more quick curiosities far afield. Many fish have a very acute sense of smell. But the champion of all the creatures is the male emperor moth. He can detect the scent of a female from 11 kilometers away, even with the wind blowing in the other direction. That beats even the dogs at their own game.

The Loudest Fish

Isn't it enough that we must listen to parrots screeching, lions roaring, mosquitoes buzzing, elephants trumpeting, raccoons shrieking, groupers grunting, and grackles chattering incessantly? Into this incredible, and actually delightful, cacophony of the natural world enters the toadfish. This creature, an ocean bottom dweller, has proven itself to be loud enough to keep San Francisco Bay houseboat dwellers awake at night and, from two feet away, is literally louder than a subway train. The male is the only culprit in this noise pollution.

Of all the creatures who have "a place in the choir," the toadfish might be the oddest and, underwater, the noisiest. It grunts and goes "boop boop" at a deafening level by using strong muscles to vibrate a heart-shaped swim bladder, an organ used for buoyancy. (Fish have, perhaps fortunately, no vocal cords.) The racket is used to attract females and in aggressive and defensive postures; the fish is very territorial. Don't mess with it.

Living in deep water most of the year, the toadfish comes into shallower water to spawn and then to guard its nest of many

eggs for several weeks, the latter baby-sitting done entirely by the male. But female readers may find this attribute only partially attractive. He is an ugly creature done in camouflage brown, and some species have dorsal spines equipped with poison glands. At about one foot long, the toadfish can sting hard if you disturb him. And in the process he may practically deafen you.

The Private Lives of Elephants

Everybody knows that elephants have wonderful floppy ears, a trunk like a garden hose, a good memory, valuable (too valuable) ivory tusks, and a leathery hide. But they are even more interesting than that.

Elephants live in matriarchal groups, the oldest female in charge. Young males go off on their own at about age thirteen, joining the males who cruise around and visit the females when the males are in musth. Often, the females in the group come into estrus within a few weeks of each other; and so the dominant bull in the area is likely to be the father of much of the next generation, making for a tightly knit group. The teenage females help raise all of the babies.

A baby elephant, which weighs about 260 pounds at birth, sometimes trips over its own trunk and also sucks it much as a child sucks its thumb. Soon, its trunk will be used also to dig for water, then curled up between its tusks for rest time. The first of its six sets of teeth enable it to quickly begin the typical elephant diet: bananas, pineapples (eaten whole with a huge crunch), oranges, mangoes, papayas, and all manner of flowers and greenery.

As an elephant grows older, even in adulthood, it plays and acts silly sometimes. The huge creatures have been seen pirouetting, curling up their tails, and tossing logs back and forth. One would rather see them this way than angry, because they could kill a human with one swipe of their trunk. They sometimes communicate among themselves with low-frequency calls, only the upper frequencies audible to us.

If an elephant becomes sick, the herd will nurse it, spraying water on it to keep it cool and brushing the flies away. If it dies, the others will bury their fellow under tree boughs and, even several years later, recognize its bones, handling them gently and quietly; they have never been observed to behave this way with the remains of any other species. An elephant's natural life span in the wild is about 65 years. Anyone who buys an item of ivory is cutting these years short for a magnificent creature.

A Sheep-Eating Bird

Sheep in New Zealand should beware of one kind of kea parrot. It will land on their warm wool, tear a hole in their flesh with its sharp beaks, then eat the fat from around their kidneys. Like almost all parrots, a noisy forest dweller that prefers to breed in caves, the kea is the only nonmonogamous parrot. There will surely be many more than two around. Sheep-eating is a relatively new habit among these birds, acquired by most of them after hanging around slaughterhouses, a testament to their cleverness, if not gourmet taste.

Insect Thermometers

On a sultry summer night, crickets can sound like bracelets jangling at a vast darkened yard party. It is only the male that one hears, though, as he scrapes his left wing cover across the filelike edges of his right wing to attract a silent female. Like most insects, crickets are more active when it is hotter (in fact, their chirps can sometimes be painfully loud if one stands too close on a hot night), and their song's tempo is a particularly good measure of the temperature. Species vary a little, but a good average method is to count chirps for fifteen seconds and then add 37. The result will be the temperature in degrees Fahrenheit. Though this cricket method is not Bureau of Standards–exact, it could well be more accurate for your yard or campsite than the weather announcer's temperature, which is probably taken several miles away.

The cricket is a curious creature. Its ears are below its knees. It can fight fiercely in defense of its burrow area. And it is considered, by some people, good luck to have one inside. As Milton said, "Far from all resort of mirth/Save the cricket on the hearth."

The chirping of grasshoppers and katydids can be slightly different from this friendly neighborhood oddity. Though long-horned grasshoppers, or katydids, create their noises the way that the cricket does, the short-horned grasshopper rubs its hind legs against its front wings. To follow the katydids (which sound like their name) as thermometers, a slightly different chirping formula is required: Count their chirps for one minute, subtract 19, divide that result by 3, and add 60. Then go to sleep. This is not a math class; it's a summer's night.

Insect Gardeners

Garden or leaf-cutter ants actually tend their own gardens in Central and South America's rain forests. Divided into dramatically different castes to do all the work, they bring home tree and plant leaves and flowers, all chopped up into tiny pieces, and place them in a fungus garden. As the fungus grows from eating the leaves—protected by a fungicide from the ant's mouth that kills its competitors and any bacteria—the ants feast on the fungus. And if a virgin queen splits off to start a new colony, she will leave with a bit of this precious fungus in her mouth.

These industrious creatures live in colonies of more than a million ants, each colony having excavated up to 44 tons of soil to make their moist, dark living space that can extend to five meters in depth. In a single night, one colony cut up and made off with every bit of leaf from an entire orange tree. Leaf-cutter ants, indeed.

The Most Numerous Visible Creature on Earth

Out of the 1 million species of known insects, beetles have the most species crawling the Earth, but if you count up individual creatures, the one that would probably win is the aphid. Many are their predators, but greater is their reproductive ability. More is the woe for us.

Aphids have virtually no defense except for their almost incredible ability to make more of themselves. Flower-fly larvae can eat them all day long. Parasitic wasps often inject their eggs into them, which soon hatch and eat up the aphids. Lacewing larvae can do in dozens of them an hour. Ladybugs treat them like delicacies. And even chickadees eat the eggs, all winter long.

The aphids have a trick up their sleeves, however: virgin birth, or parthenogenesis. When the springtime's first generation emerges from the overwintered eggs, they begin to eat and soon reproduce without mating at all. This next group does the same thing, and the next and the next and so on. Only in late summer is a single generation of males finally produced. Destined to a life so short that they have no mouths to eat with, "the boys" mate with the latest crop of females. The eggs that result are then the ones that overwinter, starting the whole process all over again.

Woe, indeed, to us. If aphids were not killed by their predators, a single female could lead to nearly 6 billion aphids in just one summer. Many would feed on our crops. Perhaps these creatures should be investigated as a filler in food for cows or dogs. (If all the insects in the world were weighed and added up, their combined weight would be about twelve times greater than the entire human population's.)

This is a record for creatures that you can see without a microscope, of course. Counting up plankton, viruses, bacteria, or tiny fungi might lead to a different champion.

The Deadliest Creature

It has probably killed more people than all the wars in history and, even today, is responsible for at least 1 million human deaths every year. It is not the lion or tiger or bear, or any terrible human military leader, or even the rat, which comes in second in this contest. Its name: the mosquito.

The diseases spread by this creature include malaria, dengue fever, yellow fever, sleeping sickness, encephalitis, filariasis (which, in turn, causes elephantiasis), and more. Thirty years ago, when malaria was actually less common than it is now, mosquitoes killed a person somewhere in the world every ten seconds with this disease alone. Mosquitoes are not about to lose their prize.

The 3,000 species of mosquitoes live mostly in the tropics but extend all the way up to the Arctic. There, vast swarms, hatched together when conditions are right, can lead to mass attacks: One such blitz, documented at some 9,000 bites a minute to the flesh of one hapless individual, could have killed from loss of blood alone.

Why all this death-dealing? A mosquito needs its "blood meal," about five milligrams of blood from an unswatted, uninterrupted bite, to make a full batch of new eggs. In the process, diseases from whatever it bit previously may be transmitted. Only the female so gorges, using protein from the blood bite to produce the eggs. Your blood is enough to turn her abdomen from black to red and swell it to perhaps twice normal size.

Each bite is a sophisticated military maneuver, requiring 50 seconds to complete and proceeding for a full two and a half minutes if the "general" is left undisturbed. Once alighted on exposed flesh, the mosquito's tiny legs probe for the most vulnerable place. Next, two lances pierce the skin, accompanied by her microscopic saw, to rip your flesh more. Then a tube is lowered through a sheath to draw the blood up, powered by two pumps in her tiny head. If allowed the time, the mosquito also injects an anticoagulant to keep the blood from clotting on the way up. It is this chemical in her saliva that seems to create the bite's itch (and it is her salivary glands from which the diseases are spread too).

Beginning about three minutes after a mosquito bite, the itch, if any, will subside, then in about an hour, it can begin again when the red welt forms. Known to be an allergic reaction, this itch varies quite a bit among people and depends also upon how often one is bitten and by which mosquito. Even the age of the creature can make a difference. The itch has not yet been well researched by scientists (and who's going to volunteer?).

People seem to vary in their attractiveness to the mosquito, with human body odors playing an as-yet undiscovered role. Several things are known, however: Mosquitoes are attracted to an elixir of carbon dioxide, ammonia, lactic acid, water vapor, and heat. In other words, the breath and body odor of a warm-blooded animal. This perfume's warmth can be detected by the creature against a background only .0005 degrees Centigrade cooler. Mosquitoes are also known to be attracted to khaki fabrics.

Of course, few people want to know how to attract them; most of us would rather use repellents instead. A very strong smell of garlic does work but would also repel people. The regular chemical repellents work, too, at least for a while, especially DEET and permethrin. They do it by making the creature lose its sensitivity to carbon dioxide, lactic acid, or both, or by acting as a toxicant. More broadly applied antimosquito possibilities now in the works include one larvicide, one bacterium, a few natural predators (a minnow, a carnivorous bladderwort plant, a tiny roundworm), and possibly the essence of African water-buck or (believe it or not) of wild chrysanthemums. Sonic re-pellents and light traps are known *not* to work. The battle is still on, against a creature who has been launching surprise attacks, successfully, for 200 million years, and is not about to beat a retreat now.

Animal Weight Lifters

The rhinoceros beetle, whose name tells it all, can lift 850 times its own weight. And there are runners-up who are also strong: The stag beetle can drag a load 120 times its weight; the ant can struggle along the sidewalk under 50 times its own weight; and even the bee can carry about 24 times its weight—honey-makings in flight.

The reason for all this strength among the insects explains why they go "crunch" when one steps on them: Their skeletons are on the outsides of their bodies. Their muscles are attached

to this exoskeleton with great leverage and mechanical efficiency. These prizewinners also tend to have more and proportionately thicker muscles than people do.

The exoskeleton, which also functions as a suit of armor and protection against desiccation, has some disadvantages for insects, though. It is somewhat insensitive to touch and cannot stretch at all. So, periodically, while it is immature at least, the insect must molt, its muscles detaching from the skeleton. It thus remains vulnerable until the outer layer of young skin hardens into a new skeleton ready for olympic weight lifting again.

Why Is the Largest Animal Probably a Female?

The female blue whale surely never counts calories, and sure enough, she is fairly likely to be bigger than her mates. Though a species where the females are larger than the males may seem unusual to us, it is actually the more common situation across the planet. Among insects, fish, reptiles, and amphibians, the female is usually bigger, while only among mammals and birds is it the male who is the big cheese.

The reasons for this "sexual size dimorphism" have been hard for scientists to figure out, but—at least among some creatures, and when there is enough food so that fertility wouldn't suffer across the female's life span—larger size gives a female an advantage in the number of young she can bear. Thus large size would be rewarded and passed on to more and more descendants, eventually allowing the females to outdistance the males, whose size was less rewarded. Or, larger mothers might have not necessarily more progeny but more fit ones, while smaller size might be an advantage to the males, allowing them to get around faster. Finally, for creatures that continue to grow, perhaps more of the females simply live longer, gaining in stature all the time. " 'Tis a puzzlement."

The Fish with the Best Directional Sense

Returning nearly 5,000 miles from its ocean feeding grounds to the stream of its birth—that one stream among thousands—the pink salmon is a miracle in motion, an iridescent arrow veering in the glint of the sea. On its voyage, it displays a combination of directional talents from an acute sense of smell to a sun compass, an eye for polarized light on overcast days, and perhaps a feel for both electric and magnetic fields. Such a variety of pathfindings is not unknown among marine species, bees, and birds but finds its zenith in the salmon.

A slightly different smell lies in each stream where salmon are spawned. As hatchlings, they receive the imprint of the distinctive scent of plants and soils dissolved in their own home water. Later, they will use this smell as the final clue, at stream's mouth.

But salmon feed far afield in the ocean, much too far away initially to smell the stream. Some Pacific salmon species swim, in fact, to the waters of Japan and Korea, others to the Yukon

River and Aleutian Islands. Atlantic salmon often go to Greenland. To return home from these far distances, they use a sense of the sun's altitude and its changing angle, a kind of inner solar compass. When the sky is overcast, the fish respond to polarized light; the sun's light is most intense in the directions away from the sun and toward it, and least intense at angles of 90 degrees to the sun. Both directional ways show talent, indeed.

Less clearly understood is the salmon's sense of electric and magnetic fields. Since salt water moving through the Earth's magnetic field creates small currents, salmon could use them as some eels and other marine species do. They could also join some bacteria, migratory birds, and honeybees, which use magnetic particles in their bodies to detect the Earth's magnetic lines of force. There are indications, not yet proven, that salmon may indeed use this clue on overcast days.

Once safely back in their home streams, the salmon swim fiercely upriver, leaping in twisted curls over waterfalls as high as 10 feet or more, and reaching speeds of 50 to 60 miles per day. They do not eat. They arrive battered, the females to lay their eggs on the riverbed, the males to fertilize them. Within a few days, their journey over, they are dead.

Bird Song

"Careless rapture," it was called by Robert Browning. "Harmonious madness," Shelley said of the skylark's song. Bird song is this and more. A bird in song—from the most glorious trill and exaltation of the lark to the homeliest chirp of the house sparrow—is a bird proclaiming rights to a breeding and feeding territory, attracting a mate, bonding with that mate, calling fellow flock members together or otherwise signaling them, or, in some cases, simply releasing energy. Though there are exceptions, the most exquisite singers are usually the plainest birds. Their song is their plumage.

There is a season to the song. Birds sing loudest and most frequently during springtime breeding—and it is the males who are advertising their real estate and sexual availability. When an intruder of the bird's own species comes close—not a stable neighbor, whose song is often recognized and discounted—the songs become more aggressive and the bird may even attack. A bird who notes no intruders anywhere near sings less. Though records are sketchy in these matters, one bobwhite was heard singing 1,403 songs in a single springtime day, and one song sparrow created 2,305. And the cowbird, moving through four octaves, has an especially unusual springtime ritual: The female raises a wing up and then down quickly to signal the songs she likes in the mate's repertoire, molding his behavior. As springtime proceeds, territories and mates are settled, but quite a few birds continue to sing to some degree to strengthen the bond; paired tropical American wrens even engage in duets, and, in a few species, the females now rival their mates' song. Some birds still sing over their eggs, fewer over their growing young—some making "landing calls" for the baby birds to recognize and thus get their mouths open—but by midsummer almost all have stopped their song. One of the few, the plain brown oven bird, does sing all summer long. Small birds like finches and vireos, too, will keep their "mobbing calls" in any season, a signal for their fellows to assemble to help drive away a predator. By fall, quite a few species begin to sing again, though in weaker and shorter songs; a very few, like the American goldfinch, have a "social song" to help the flock gather for migration. Among migrants, winter often shows a bloom in female liberation. To set up their own feeding territories once they arrive in the South, female birds can sing as aggressively as any male in springtime. And birds who overwinter in the North, like the cardinal, often sing a bit on a cold but sunny day, bright red against evergreen and white snow. He could be thinking about springtime territory already.

During the course of a single spring day, birds' singing patterns vary considerably. Robins prefer to sing at dawn, a bit in the afternoon, and then again at dusk; while whippoorwills, truly crepuscular, sing just before dawn and just before dark. Sunset is the time of the thrush, usually called the most beautiful songster of all. Vireos sing throughout the day. Night singing is common among loons and great horned owls but is also done by daytime birds such as the mockingbird and nightingale. In addition, any bird will sing anytime it needs to defend its territory. And the mockingbird can imitate a jazz record or a dog barking at any point.

Some birds have quite a panoply of substitutes for true song, all in the name of communication with their fellows. Woodpeckers drum their bills, white storks clatter their jaws, turkeys click, one kind of sparrow "hiccups," some grouse make drumming noises with their feathers, Canada geese conduct honk "conversations" with the younger birds during fall practice flights, and domestic pigeons clap their wings (for a long or short time depending upon how long a flight they plan to make). The cuckoo uses calls for trickery—the male calls loudly enough for a nesting bird of another species to decide to chase it away, thus allowing the female cuckoo to lay their egg in that bird's nest. And baby birds of the pheasant and partridge types "click" to each other, even from within their eggs, before hatching, a signal that seems to coordinate hatching times.

All this music is sometimes known instinctively and sometimes learned, depending upon the species. Some birds are practically born singing, even if they are raised in isolation. Baby chickens who receive brain transplants from quails have been known to come out of surgery sounding just like quails. Chaffinches and western meadowlarks can produce an imperfect song without practice, a perfect one with exposure to adults of their species. And canaries require a full eight to nine months of practice—with the development of extra brain cells—and much soft, bedtime "babbling" to achieve their glorious song.

As a by-product of either learning or inheritance, the songs of some bird species sound as if in dialect. The songs can be discernibly and consistently different even when the bird groups are only 17 kilometers apart, in some cases. European wrens sound very different on and off the British mainland. And a cardinal who lives in Texas sings a different song from a cardinal in Minnesota, even in its ecstasy.

In Praise of the Cockroach

Very little good can be said of the cockroach. But consider this: The cockroach is truly good for the economy. Every year Americans spend nearly a billion dollars on extermination, professional and amateur, for home and work place, of this tricky and unattractive insect. And there is no end in sight—we are investing in a 320-million-year-old creature, and one that has kept essentially the same almost indomitable form

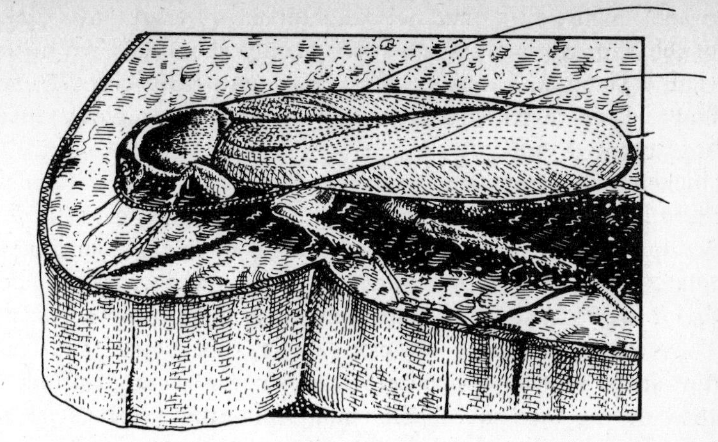

for at least 300 million of those years. Cockroaches serve an additional, though small, economic role in the chicken feed and fish bait business, as feed and bait. Think of the jobs provided in all of these industries.

Indirectly, too, they add to the incomes of doctors. Richly active, they spread salmonella food poisoning, cholera, dysentery, plague, hepatitis, and more, all via their excrement and vomit and through direct contact. According to some scientists, many food allergies are actually caused by cockroach excrement either left on the food or airborne with the household dust.

And we are not finished yet with their economic impact. One kind of cockroach, the Madagascar, is often kept as a pet by entomologists. Doubtless they buy a box of cornflakes or crackers for each of them now and then, since the insect is a hefty three inches long. This surely swells the economy further. These scientists usually also have two of the pet insects so that they can watch them duel or mate. (The males can butt heads and horns for half an hour at a time, hissing very loudly. The female will hiss more softly but also wheezes and rasps for one's entertainment.)

Let's call it an even billion dollars, of economic boost every year in America, from this insect—and quit while we're ahead.

Well, in fact, they are. The cockroach has already produced on this planet an estimated 1.5 billion *generations,* each one of which can easily run to more than 10,000 individuals descended from a single mother and her daughters every year. Fortunately, of the approximately 4,000 species, only about 550 live in the United States and only seven of these are truly pests. Everybody's favorite cockroach pet, described above, lives *only* in Madagascar.

The generic American cockroach lives well. It can eat almost anything, live almost anywhere, and defend itself expertly. An intrepid scientist who examined the gut of one of them found bits of "boiled potatoes, vegetables, cereal, dough, dead and diseased cockroaches, chocolate, honey, butter, vaseline, bread flour, sugar, feather, wool, cloth fibers, shoe polish, and book bindings." They also like beer (and so can be found drunk in the morning—and squashable—on a beer-soaked paper towel left out overnight).

Failing these delicacies, the cockroach can live for three months without any food at all and for one month without water. They can also tolerate freezing for two days and much more radiation than we can. Their body walls are even waterproof.

Versatility is the name of their game in habitat, too. Most enjoy the tropics and their abundance of dead vegetation, but they will live anywhere except for polar regions. One very small species even lives on the backs of certain fungus-growing ants, inside the ant colony. In our cities, they live inside food containers, also chairs, even radios—anywhere that is reasonably moist and warm. And where you see one, there are more.

But, everywhere, cockroaches tend to dislike light, noise, and sudden movements of air, all of which bespeak danger. Because of these proclivities they have perfected extraordinary defenses against being squashed. From each cockroach's body protrude very sensitive sensory hairs, about 220 of them in an adult. Even the gentlest puff of air bends these hairs, sending a simple but effective neural signal that then makes the insect run. This escape is like a bad miracle, beginning in five-hundredths of a second; thus they can easily outwit a human strike by shoe or rolled newspaper. Light and noise sometimes make the cockroach flee this way, too, though at other times these conditions make it stop in its tracks to take advantage of camouflage. No wonder they are so difficult to exterminate.

Often a creature that has survived many millions of years the way that the cockroach has is highly specialized, having evolved to essentially one simple ecological niche. It is common for such a creature to live in the ocean, where the environment has been more consistent over the millennia than the land's has. But the cockroach has taken the other approach—that of a generalist. Whether it eats leaves or soap or cotton, lives in the jungle in Brazil or inside a flute in Duluth, it always contributes well to the economy.

The Insect That Outweighs a Mouse

Try to swat this insect and you might bruise your hand. The giant weta is probably the heaviest insect on earth, heavier even than the Goliath beetle. Many adults tip the scales at two and a half ounces and are about four inches long.

Three species of this huge cricketlike creature—all of them fat and shiny and gentle—live in the wilds of New Zealand. Because of their size, they can neither fly nor jump and spend their entire lives on or near a single tauwhinu bush. They are, fortunately, both vegetarian and nocturnal. Giant wetas grow to maturity in 18 months, mate for another half year, lay a few hundred eggs, and then die before the new little giant wetas hatch. Once in danger of extinction because rats were introduced to their islands, they are now safe on rat-free Maud Island in the New Zealand chain, to which they were moved—gingerly—by a local ecologist.

New Zealand, ecologically isolated for most of the last 100 million years, is home to many especially primitive and odd species of animals, as readers of this book will probably notice. They have had a chance to evolve in their own ways there in a world crucially free of snakes, crocodiles, and land mammals as predators. Insect species tend to be particularly individualistic, in New Zealand and elsewhere; some have been around for nearly 450 million years of subtle adaptations.

But the giant weta is a special curiosity because of its size. Most other insects are less than two inches long, and this constraint usually results from their breathing apparatus. Insects lack lungs, instead using a system of tracheal tubes that allows oxygen to move directly to body cells from openings on the outsides of their bodies. This air action is fast and efficient but works best with short tubes. Though some insects have expandable tubes and others use their muscles to increase air circulation, even these refinements have their size limits. The longest stick insects and broadest giant moths measure no longer than ten to twelve inches. And though there were once larger dragonflies in prehistoric times, we will probably not see an eagle-sized bumblebee or a man-eating cockroach any time soon to outdo the giant and gentle weta.

Ant Adaptations

There is an ant species for practically every small purpose on Earth, and most are innocuous from the human perspective. Three species stand out, though, in the significant interactions they can have with us.

Get out of the way of the fire ant, the nastiest ant to humans. Its sting feels like a burn for about five minutes, then turns to a welt, and next a smaller pustule. This all takes a full week's worth of itching and may leave a scar too. And, worse, these ants will climb all over you, each worker stinging more than once. These tiny reddish-brown creatures are rapid colonizers; when a queen flies off to start another colony, she does a fiercely good job—by the end of four years the new mound nest can be home sweet home to 250,000 fire ants. And, worse than that, they are evolving to have more than one resident and breeding queen per mound, sometimes as many as 500 queens. Their homes are all over the southern United States, and spreading. Although the ant's normal diet is other insects, spiders, and earthworms, they sample crops such as soybeans, citrus trees, corn, okra, eggplant, and birds' eggs too. Their mounds can be in—and can damage—airport runway lights and traffic signals. They have even been known to kill people by multiple stings.

The fiercest ant, though, is probably the marauder ant of the Asian tropics. The soldiers, one of their castes, are about three-quarters of an inch long, yet one of them can crush a two-inch centipede in its jaws. This is after the hapless prey has been pinned down by some of the one-tenth-of-an-inch minors, another of their castes. Spiders, cockroaches, crickets, and scorpions can meet a similar fate, as do many more seeds and fallen fruits. These ants live in groups of hundreds of thousands and advance in swarm raids across the jungle floor, a bit like their distant cousins the army ants, which advance in columns like armies.

On the other side of the ledger, the honey ant is truly sweet, at least from our perspective. Some of these ants fill their abdomens with nectar until they become as big as blueberries, then go back to the nest to offer fast-food on demand to their fellows. Dig some up as the Indians used to do and eat them like grapes.

The ant kingdom, which includes nearly 15,000 species and dates back 100 million years, has untold curiosities. Some ants herd other insects (particularly treehoppers and aphids, who constantly excrete sugar for the ants), some enslave each other, and some use their own larvae as weaving shuttles (thus creating large nests in the rain forests of Asia and Africa). Each ant has about a half million nerve cells in its brain, which appears to be enough to make some clever critters.

How Do Silkworms Make Silk?

Much like the glinting silk of a spider's web is the silk knit by a silkworm. This unusual insect is a type of caterpillar and makes silk to create its cocoon. The silkworm builds its soft home in about two days, using a full mile of silken thread, all of it spun out of silk glands that may add up to more than one-fourth of its body weight and open to a spinneret on its mouth. To weave, it moves in an unceasing, swaying figure eight as the silken saliva, first liquid, quickly hardens.

No bigger than a grain of rice at hatching, the silkworm molts four times to reach its spinning size, about that of a little finger. Fed on mulberry leaves, this insect larva has become a kind of domestic animal in various parts of the Orient, where it does much of its delicate work.

Silk spinning is also done by some beetles, the larvae of many caddisflies, various other insects, and all caterpillars. The latter

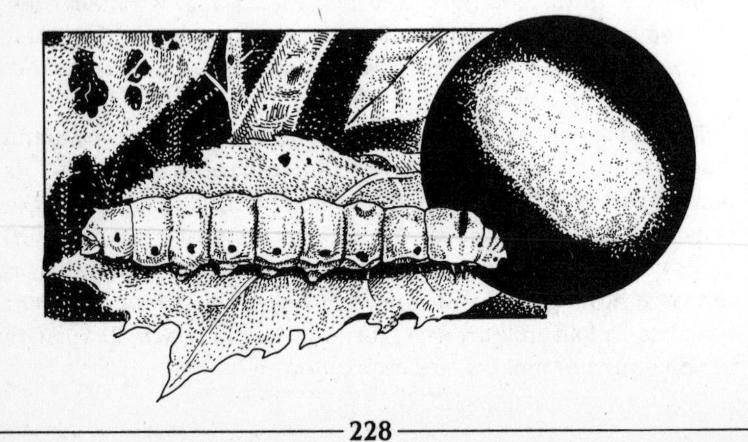

use their silk to secure a foothold on a leaf, to mark their path, to escape by "rope," to line tunnels, to create cocoons, and to make webs. Some spin in different colors depending upon what they have recently eaten. But you can't feed a silkworm roses to get a ready-made magenta blouse.

Expert Parasites

Bloodsucking leeches and tapeworms are only two of the many kinds of parasites. Parasitism is found among all groups of plants and animals. And it is probably as important as predatory behavior for balancing the numbers of living things on Earth (especially when one considers that all viruses are parasites too).

The phenomenon of parasitism often seems to evolve toward a simpler, but related relationship called commensalism—which comes from the Latin words for "eating at the same table." The sea anemone and the clown fish are truly commensal, the fish living among the sea anemone's tentacles and sharing its food. So are remora fish and the sharks to which they attach, then loosen themselves periodically, to find tiny fragments left over from the big fish's meals.

True parasitism is less neutral, however. Instead, it is a relationship in which one side benefits and either gives nothing in return or actually harms its host to advance its own life. Fish are typically cluttered with small parasites, such as the fish louse with its hooked claws and the bloodsucking leeches. More serious are the species of lampreys that fasten onto fish with wide, jagged mouths, inject a digestive enzyme, and slowly eat the fish's flesh. Aphids are parasitized by wasp larvae and, under that pressure, have evolved to kill themselves sometimes to damage the wasp's chances (thereby benefiting their fellows) or, alternatively, to choose a protective place to benefit the wasp in a long overwintering. Vampire bats attach themselves and lap up animal blood with their tongues. The European cuckoo and the American cowbird lay their eggs in other birds' nests, a form of "social parasitism." Sand wasps sting caterpillars into paralysis, then drag them to their holes for the wasps' larvae to eat. And ticks and fleas infest dogs and humans. Among the eeriest parasites from our perspective are the hookworm and trichina worm (transmitted in uncooked pork), one nematode that can cause elephantiasis, and a type of tapeworm that can grow to 30 feet long in our intestines.

Termites, which lack the enzymes to digest wood (their main food), can do so with the help of microorganisms stationed in their intestines. These tiny organisms use some of the food and leave the rest for their hosts. Some 26 species of fish remove parasites from other fish, meeting all of their own food needs in the process. And some ants carefully tend the broods of young treehoppers and other insects that produce honeydew, one of the ants' favorite delicacies; this baby-sitting stimulates treehopper parents to produce another brood, which proves quite useful to the ants. This type of relationship is called mutualism, a situation in which both sides have evolved to benefit each other.

Evolution must have stretched long and taken odd avenues to arrive at adaptations like these. It is a sign that ample chance has been capitalized upon well.

A Fish in the Desert

In Death Valley swims, yes, swims, the tiny desert pupfish, sometimes in springs and streams and marshes no bigger than a bathtub. This water is all that is left of the vast Lake Manly, which, during the last ice age, covered a large area of what is now called Death Valley. Some twenty different populations of the pupfish have been found here, each in its tiny oasis and by now divided into four different species varying slightly but clearly in their shapes and splotched markings. None is larger than about one and a half inches long at adulthood.

Any desert, any time, is a hard place for a fish, but summer is especially hard. Even though the pupfish can tolerate a temperature range of about 70 degrees Fahrenheit and drastic changes in salinity, millions die each year as their habitats heat up and partly evaporate, becoming much saltier in the process. A population in a single place can fluctuate more than a hundredfold between the later end of the killing summer and the end of the mating spring. A flexible fish, indeed.

Other organisms can survive extreme desert conditions, too, by even more extreme measures. Brine shrimp can dry out almost to death, even for decades, awaiting moisture. And some tiny pond dwellers called rotifers and tardigrades, revived in a laboratory, briefly, when they were accidently watered after 120 years.

The Irritated Oyster

Oysters do not yowl or even whine when irritated inside their shells. Instead, they make pearls. Every lustrous pearl is evidence of irritation caused an oyster by a grain of sand, a small egg of a parasite, or even, perhaps, a bit of excess oyster waste caught within its shell by accident. Any of these cause the creature to smooth over the area with the same substance that makes the "mother-of-pearl" inner gloss in many other shells. As layers are built up on all sides of the offending particle, a precious pearl is formed. The process can take several years.

Besides the pearl oyster—which, unlike the common edible oyster, lives on the bottom of tropical oceans—pearls are also made by fresh-water mussels; though they are rarely as exquisite, they do come in tones of red and pink.

Carpenter of the Sea

The sawfish, aptly named, actually sports a saw—and saws other fishes to death in order to eat them. It looks a bit like a shark. Approaching a school of fish, it lashes its double-notched saw back and forth, sometimes even into the ocean floor to discover hidden prey, then gorges on the victims (leaving a few for later passers-by). It will also attack large fish, even whales, eating the skin and entrails torn out in the puncture wound. These flat-headed, fetching carpenters are bottom-dwellers in tropical and subtropical seas but have been known to swim up rivers a bit and saw into human swimmers in India.

Why Geese Fly in a "V"

The fluid "V" of geese, windswept across a background of sky, is not only dramatic but may be energy efficient. As each bird flaps its wings, it creates two whirling eddies of air. The rising part of each spreading vortex produces aerodynamic lift that pushes up on birds following behind. This "air wake" works much as a water wake does, since both are made by turbulence. All stand to benefit but the very front bird in the "V." Research also indicates another possible reason for flying in a "V" or partial "V" is to keep other birds in sight. Geese eyes are set on the sides of their heads and can't rotate in their sockets the way people's can. To keep all the other members of the flock in sight while still looking forward, a staggered formation such as a "V" works well. Besides geese, other large and gregarious birds fly this way, including species of swans, cormorants, godwits, ibises, ducks, gulls, and other shorebirds.

The Biggest Rodent

Tipping the scales at up to 200 pounds, the capybara weighs in as the world's largest rodent. Luckily, it is vegetarian. A capybara looks a bit like an enlarged guinea pig (to which it is related) with its large head, square muzzle and short, rounded ears. Its body, up to three and a half feet long, is covered with sparse, reddish brown or gray fur. An inhabitant of the densely vegetated areas around the ponds, marshes, and streams of South America, the capybara eats grasses and, sometimes, water plants, grains, melons, and squash, padding up to much of its dinner on partly webbed feet. It also has large teeth. It rests in thick patches of vegetation during the hot part of the day, then is active in the morning and evening. If bothered by people, however, it may become nocturnal. On land, the capybara will run like a horse if startled or may head for water. It is a capable swimmer, and dives, swims underwater, and may even take refuge in floating vegetation, leaving only its nostrils in the air. Both the nostrils and its tiny ears are self-sealing to keep the water out. Capybaras usually live in groups of about 20 or so, in which a strict social hierarchy is maintained. Their sound is a "tweetle-tweet." Capybaras are still fairly common, although hunting has decreased their numbers in recent years.

Fish Talk

Fish have a lot to say—and a lot of ways to say it. They communicate with the "lateral line" that stripes their sides, make noises from grunts to barks, glow in Technicolor, flash light, use body movements, and employ blasts of electricity.

Evidence of this is seen in their schooling. A vast scarf of fish, moving in rhythm and without collisions, can indeed billow over a seamount and ripple through a splash of undersea sunlight as though the school were a single fish. It looks impossible to do, especially when the school turns in unison. This synchrony of swimming is accomplished by means of an odd sensory organ called the lateral line—half touch and half hearing. Although all fish have this sensitive strip, running from near their heads to near their tails on both sides, it varies in its sensitivity and only about 4,000 species use it for schooling.

The lateral line is sensitive to the "chatter" of minute electrical currents, the low-frequency vibrations and pressure waves created by movement through water. Since each fish "hears" them, each can adjust to maintain the separation optimal for its species. The system seems to work most efficiently when a school is moving relatively fast, and it can be relied upon at night as well as day. The school can change direction in both deep and shallow water so easily because one edge of the school can lead as well as another—there are no natural leaders and followers among fish.

Schooling seems to have evolved as a protective device. Each fish is, individually, safer in a school; this helps it survive so its ability to use the lateral line for schooling can be transmitted to many offspring, maintaining the skill in the population. An undulating mass of fish seems to truly confuse some predators: A school of minnows, each tiny but adding up to a mass nearly 100 feet in diameter, can look daunting, and anchovies can even tighten into a ball to deter attackers. Schooling in a growing group of fish seems to develop gradually as fry meets fry, and a group will spin off to form a separate school once about half the fish become significantly larger or smaller than the others.

Those fish that use their lateral lines to "converse" with prey rather than to share the old school tie are usually especially sensitive along these lines. Some deep-sea fish can use their lines to find a prey fish up to 60 feet away in complete darkness. The whalefish, a blind inhabitant of 18,000 feet down, has a gigantic hollow tube with large pores for its line. And frogs use a somewhat similar organ, lateral buds, to detect the waves made by insects.

Lateral line is very quiet talk indeed, however, compared with the noises fish make by turning their swim bladders (used to maintain buoyancy) into bellows or flapping their body parts. One can hear grunts, clicks, barks, snaps, and groans in the so-called quiet sea, all of which are sounds used to attract mates or warn rivals away. The small male damselfish chirps for a date to come to his nest. Large groupers try to boom away intruders. And drum fish and squirrel fish are especially noisy at dawn and dusk, like many birds.

Changing color is another communication trick fish employ, though it is more often used for camouflage than for talking. Cichlids and others often change for mating and territorial battles; damselfish turn blue in response to aggression; and one fish from India fights solely by flashing colors. Other fish use color to highlight those body areas in need of cleaning by smaller fish. Fish can usually change their colors in a matter of seconds; they accomplish it with pigments and reflective tissue in a layer of cells just beneath their scales. Emitting light, as opposed to changing color, is common.

The movements of fishes' bodies are also a form of communication. Some angelfish will lift a fin as a signal, and gray reef sharks can turn themselves into wild and twisting "S's" to warn others away. Still others use an open mouth to show aggression.

Finally, electricity is yet another way of conveying messages, as about 700 species can create electrical fields. Some sharks use receptors on their snouts to sense these fields—some as much as 25 million times weaker than what we could detect—a practice known to be useful for finding hidden prey and perhaps for other communication. Others, such as knifefish, can recognize their fellows by their electrical signals.

The Giraffe's Reach

The long-necked giraffe, whose name comes from the Arabic word "zarafah," is spectacularly well adapted to eating very high leaves, and thus fills a unique niche for mammals. It has just seven neck vertebrae, the same as most mammals, though its own are much longer. The famous giraffe neck must have developed gradually from these vertebrae, as generation after generation of the animals with the longest necks ate well and so were fit and healthy enough to have many progeny, thus passing this genetic advantage on to them.

Once thought to be one of the wonders of the animal world—and even paraded through the Coliseum by Julius Caesar—giraffes are physiologically extreme. They have bulgy eyes (and use them to see well in all directions), hairy lips, a tongue as long as 18 inches (convenient for curling around leaves), and a 25 pound heart for pumping blood to their extremities. They often eat more than 125 pounds of vegetation every day. At 15 to 18 feet tall, their heads are so far above their hearts and lungs that pumping blood to the brain also requires a very high blood pressure—an average of 260/160 (compared to an average of

120/80 for humans)—and very deep breathing, at the rate of more than 20 breaths per minute, to move oxygenated blood up the neck to the head. When they bend down to get a drink of water, which they can go days without, special valves close to prevent a huge rush of blood to their brains, and their legs spread as if to split. They are quiet creatures, producing only little grunts and cries, but seem also to "converse" with ultra high-pitched sounds.

A lesser-known stretcher is the gerenuk, sometimes called the "giraffe antelope" because it stands high on its hind legs to eat leaves as well.

The Barking Deer

The little brown Reeve's Muntjac, usually shy, barks in mating season or when danger is near. This deer is about the size of a dog too, though it can jump higher fences than most canines. Beloved by Britishers in the countryside, they leave food out for it (but probably not in Fido's bowl).

Animals That Look Like Flowers but Sting Like Bees

If the sea is a vast garden of life, then the most spectacular "flowers" in the garden are the sea anemones. These round, colorful ornaments of the intertidal and subtidal world are not really flowers or even plants, but simple animals, relatives of the corals and jellyfish. They come in a lavish variety of colors, from pink to white to gorgeous blue-green. No marine museum would be complete without a specimen or two. Yet despite their beauty and simplicity, sea anemones display an astounding repertoire of behaviors and abilities all directed toward helping them survive in the deceptively peaceable-looking kingdom beneath the waves.

More complex than a sponge, the sea anemone's body is a tube anchored to its substrate by a flat, fleshy pedal disk. The tube is the animal's gut. At the top sits the mouth, crowned by a ring of soft tentacles resembling the petals of a flower. The tentacles catch the zooplankton eaten by some species, or the small crustaceans, fish, and mollusks eaten by others. Once captured, the prey is thrust down through the mouth into the gut and digested. (Anemones know no fear when it comes to feeding; one individual was even photographed ingesting a small shark.)

Anemones have no brain or other centralized nervous system, yet they can recognize other anemones by sex or other features. For example, some anemones reproduce by pinching themselves off at the base and splitting into two smaller, genetically identical clones. An anemone that makes contact with another can tell whether the other is a clone of itself or an unrelated individual. Furthermore, an anemone can learn to recognize neighbor anemones that have shared its living space for a while, and tell them apart from newcomers invading its territory.

For most anemones, life is a constant battle to gain a foothold where food is adequate and defend their space against others with the same goal. A coral reef-dwelling anemone may squeeze in between living blocks of coral, fastening onto a piece of dead coral. Anemones of the coastal intertidal regions compete for space on rocks near a mussel bed, where they can grab hapless mussels knocked loose by the surging tide. Anemones attached to wharf pilings are likely to share their space with barnacles. In any case, another organism is always ready to move in should an anemone relinquish its piece of real estate.

Anemones come well-equipped for the struggle. Their tentacles carry deadly little packets called nematocysts, which can be discharged to capture prey or attack a foe. One type of nematocyst releases a barb that sticks in the target's flesh and paralyzes it with poison; another kind, called a spirocyst, shoots out sticky threads useful in snaring a mollusk, crustacean, or other shelled prey.

In "deciding" whether to shoot its nematocysts, at least one small anemone seems to use primitive senses of smell and hearing. Near the cells that fire nematocysts lie other cells that can pick up the "scent" of prey. When they do, they appear to tune a third type of cell to "hear" the prey. Nematocysts are most likely to be fired when those cells hear the frequency to which they're tuned. In the laboratory, the smell of prey resulted in an apparent retuning from 55 to 5 hertz. Five hertz is the frequency churned out by the anemone's favorite prey—a tiny shrimp—as it swims. Thus the aroma of food prepares the anemone to better hear its prey approach and so not miss the chance to catch it.

Some anemones carry a few "fighting" tentacles in addition to their crown of feeding tentacles. The fighting variety is much longer and can reach out beyond the mouth area to attack an intruder. Additional to this are the "sweeper" tentacles of some coral-dwelling anemones, which may be ten times longer than feeding tentacles and loaded with nematocysts. An anemone crowded by coral may develop and inflate sweepers, then swing them out to destroy the offending coral. However, some coral can develop their own sweeper tentacles in response and eventually defeat the anemone.

One anemone group develops special weapons for fighting other anemones. The weapons—white, inflatable sacs that grow below the tentacles—contain poison capable of digesting another anemone's flesh. When two anemones of this group make contact, they inflate their sacs, pull back their tentacles, and swing the sacs at each other in a graceful arcing motion. After some bruising, one anemone, usually the smaller one, will retract its tentacles and sacs and slowly back away.

With all these opportunities for combat, it is little wonder that anemones have learned to recognize established neighbors. It saves them from fighting the same anemones constantly, perhaps at the cost of ignoring a real threat from a newcomer.

Anemones adjust not only to each other, but to conditions of tide and surf. A great green anemone of the North American

Pacific coast may extend 20 centimeters across and 7 high in a protected place where water flow is fairly gentle, but squash itself to a pancake only 2.5 centimeters tall in an exposed channel where the flow is fast. Such flexibility comes in handy for dealing with strong currents and crashing waves. But not all anemones live near shore; some are found as deep as 10,000 meters.

Besides cloning and sexual reproduction, in which eggs and sperm are shot out through the mouth and combine in the water, anemones display some surprising methods of making babies: Some split horizontally after blossoming a new crown of tentacles halfway up their bodies, and one species even swallows its tentacles and nurtures them as babies.

Anemones make fearsome enemies, but not for everyone. In the Red Sea and the Indo-Pacific, clown fish move with impunity through the tentacles of their host anemones, perhaps the result of a protective mucous coating to ward off the anemone's stings. The clown fish gains the anemone's protection from predators and also picks up some tidbits of the anemone's food. In return, the fish helps repel intruders and probably gives the anemone a free cleaning service.

Hermit crabs also find safety in the tentacles of anemones. In the Mediterranean, anemones shelter crabs from marauding octopi, while a Florida hermit crab escapes the clutches of larger crabs in the tentacles of its anemone host. One anemone attaches itself to a hermit crab shell and hangs on for the ride. It pays its passage by secreting a horny membrane that extends the size of the crab's shell and so lets the growing crab keep its mobile home a little longer. Another anemone takes it one step further by building a whole shell for its crab companion. The anemone does such a good job that one biologist, happening upon such a setup, discarded both anemone and crab and described the perfect gold shell as a new snail species.

On the whole, anemones pose little threat to waders. The usual result of stepping on one is local itching or swelling, although the venom of a few species can cause fever, chills, abdominal pain, or vomiting. Some South Pacific species are cooked as food, and eating an uncooked serving of an especially venomous species was once a method of suicide in Samoa. But anemones are definitely not our enemies. With their spectacular colors and intriguing adaptations to the rough-and-tumble life of the ocean, they deserve to be counted among our friends.

The Grim Reaper of Crops

Every year, fungi destroy billions of dollars worth of crops by causing diseases in the growing plant and then by spoiling the stored food. Millions of dollars are spent annually to try to do them in, while every year they destroy enough food to feed 300 million people. There are a full 40,000 different species of these creatures. It was a fungus that caused Ireland's potato famine, a crop loss large enough to force many Irish to leave their country. Evicted by fungi.

Fungi can also gain the upper hand with humans, such as those suffering from AIDS and cancer, and with transplant patients, all of whose immune systems have been suppressed. Our bodies must be constantly alert against them.

Conversely, fungi are also used to make medicines such as penicillin and cephalosporin, to make foods such as bread, to ferment beer and wine, and to create food additives such as citric acid. And without them, dead vegetation would take much longer to decay, the process that transforms death into food for green plants, fungi's most powerful effort of all.

A Feathered Ruminant

Along tropical South American streams lives the hoatzin (ho-what-sin), a bird with a crest like a spiked punk hairdo, a marked clumsiness in the air, but superb climbing and swimming skills. Weighing nearly two pounds, the pheasant-sized hoatzin is brown, with light spots on the back and tan or ruddy breast feathers. Its red eyes are set off by blue facial skin. In the air, its ungainly flight usually ends in a crash landing, but in the water it is the picture of grace. When danger approaches, baby hoatzins jump from their nests and plummet as far as 20 feet to the water below. Swimming speedily to the thick vegetation at the water's edge, they wait until the coast is clear, then pull themselves up using a holdover from their reptilian ancestry: a pair of sharply curved claws on each wing, set on the second and third "fingers" of the reptilian "hand." The wing claws, which persist for the first three months of life, mark the hoatzin as a primitive bird similar in this respect to archae-

opteryx, the first known bird. Several other birds have rudimentary wing claws, but the hoatzin is one of only a few species that actually use them for climbing. Hoatzin chicks are tended by an extended family that includes older siblings as well as parents. Young and adults both live on foliage, but unlike any other bird, they digest largely through fermentation in the front part of the gut much as cows do. The crop and lower esophagus have been enlarged and modified into the main microbial fermentation chambers, leaving less room for the keel of the breastbone, where flight muscles attach; this is why the hoatzin is a weak flyer. The increased attention and protection the young receive from having several family members on nest duty helps increase their chances of surviving attacks by arboreal predators, but does little to protect them from piranhas, caimans (South American alligators), and other watery enemies they may meet while escaping the threat from above.

Animal Play

A dog flings a rubber ball into the air, then chases, catches, and shakes it again. Ravens swoop down to retrieve snow chunks in mid-air, then do double rolls from on high as gravity draws them down. Chimpanzees çareen after each other around broad-based trees, one holding a bit of booty just out of reach. Plump manatees turn somersaults and headstands with no serious purpose. Dolphins ride the bow waves of boats together, a sleek gray symphony of motion. Wolf puppies can chase, tackle each other, and scramble for 45 minutes at a stretch. Right whales raise their tails out of the water at the right angle and sail along together like boats. Though animal play like this—action without an immediate need—is in part skill rehearsal, it is also more than that. Play has, as an additional purpose, the building of bonds within the group and the creation of flexible responses to different life situations. True play is most common among intelligent social animals, especially those with wide-ranging styles of life. Its obvious forms may be more common among males in some species. And, in a certain few cases, including that of humans, it is found even among adults.

The rehearsal aspect of play, learning skills and strengths for later life tasks, is perhaps the most easily discerned. A kitten pounces on a ball as it will later pounce upon a mouse. Coyote pups tussle, then pin each other to the ground, one finally rolling belly-up to halt the fracas. Otters slide down muddy banks, vying to reach the water first. This kind of play can be solitary or sociable, involve objects, or be all three in stages, depending on the animal's needs. Activities most often rehearsed are fighting, catching prey, and escaping from a predator, with all roles taken by comembers of the species. In rehearsal play, animals practice the means but do not achieve the ends. The tone of the activity is vastly more casual and more repetitive than the "real thing" would be.

Social bonding is another function of play, though one found almost exclusively among social animals. Lion cubs trust each other not to inflict great pain in play, though they play-fight hard. Rhesus monkeys seem to enjoy the physical contact but will break up a play session if both animals are not of essentially equal strength. Older monkeys, with greater ability to hurt each other, learn to be yet more cautious about playing. The animals seem to learn "fair play"—how to calibrate and control their aggression when required to do so—while retaining the advantage of mutual touch. In the case of the rhesus monkeys, the males of which leave the troop when they become sexually mature, finding a pal to sally forth with later may provide an additional advantage to childhood play. In this species at least, the males do more playing than the females do.

Play seems to increase flexibility, too, especially among animals that range over varied terrain or that must be adaptable in other ways. Monkeys scrambling down cliffs while tussling are using two skills at once. Even adult male polar bears, ashore when the ice melts and waiting for the females and young to return from inland, wrestle and chase each other, seemingly to keep in shape. The animals that must adapt the most, to each other and to their terrain, seem to play the most—species such as timber wolves and killer whales (both group hunters) and dolphins, elephants (whose female adults play with great vivacity), parrots, and crows—all wide-ranging species. Crows, part of the most intelligent and playful bird family, have actually been seen doing stunts for an audience of other crows: One picked up a small rock and got chased, and another hung upside-down from a branch by one foot, then alternated feet. The species in which adults play also seem to be those in which both cooperation and competition with those of one's own species are common and necessary. The play of human adults—at both golf course and party—illustrates this function of play.

Human play may well have boosted our evolution: Playful babbling, rhyming, and singing probably honed our increasing speech skills, while games with primitive balls likely helped our hunting skills to develop. And acting out and telling stories around campfires may have aided us in creating the flexible roles our human cultures exhibit.

Killer of the Sea

Slicing the water like a torpedo, its body a study in power and speed, the killer whale commands respect in every ocean of the world. Even the lordly blue whale is powerless to defend itself against an attack by a pod of killers. To seals, sea lions, dolphins, and other small marine mammals, the appearance of a killer usually means sudden death. The whalers and sailors who dubbed killer whales "wolves of the sea" had good reason to do so.

Ironically, the nickname turns out to be more apt than the old seafarers could have imagined. Like their canine counterparts, killer whales hunt in packs and thus manage to take prey much larger than themselves. And on closer observation they are revealed, like wolves, to have a tightly knit social structure, a playful side, and a nearly spotless record in their contacts with humans. Wolves of sea and land both have suffered misunderstanding and unjust hatred because their hunting skill has obscured their intelligence and individuality. Off the coast of Washington State and British Columbia, killers have treated whale watchers to shows undreamed-of just a few decades ago. From playing with seaweed to babysitting each other's calves, the whales' behavior has astonished at every turn. The chilly waters have yielded a new view of the killer as a complex animal, at once tender to its own kind and mercilessly efficient toward its prey.

With their six-foot dorsal fins and distinctive markings, killer whales are unmistakable even to the greenest observer. Black dorsally and white ventrally, each has a white oval spot behind the eye. Just behind the dorsal fin lies a gray patch called the "saddle." Absent in some whales, it varies in size and shape and

can help identify individuals. Each jaw usually holds 12 conical teeth, although the number varies between 10 and 13. As the largest member of the dolphin family, killers reach lengths of up to 30 feet, with males a little bigger than females. The heaviest weigh in at about 9 tons.

Feeding a bulk like that requires a voracious appetite. In 1862, a 22-foot specimen was found with 13 dolphins and 14 seals in its stomach. Relatively few killers have been found with empty stomachs, a testimony to their superb hunting skills. They take fish as well as mammals; in Puget Sound, for example, they dine mostly on salmon. Whatever they eat, they devour between 5 and 10 percent of their weight every day.

Killer whale pods seem to be dominated by females. At eight or nine years of age the whales begin to reproduce, an activity that induces males to leave their mother's pod temporarily while they visit females in other pods. Pregnancy lasts 16 months, and a typical newborn measures eight feet and weighs 800 pounds. Besides the mother, calfless females called "aunties" may play an important role in a youngster's upbringing. Males occasionally help too. Researchers off the Pacific Northwestern U.S. coast once watched a young male babysit three calves, who proceeded to cavort in a most unmanageable manner. When the mother returned, the babysitter hightailed it for the outer limits of the pod, as far away from the calves as he could get.

The whales once sprang a real surprise on an observer watching them play. Taking advantage of the gas-filled bulbs that help some kelp species stay afloat, they grasped the kelp in their mouths, dived down, and let go. The kelp rocketed to the surface, apparently to the whales' amusement. Kelp shooting thus joins the list of killer whale games, along with breaching, somersaulting, and tail lobbing.

Pods keep to themselves most of the time, but a meeting of pods occasions a strange ritual, observed many times in the waters off the Washington and British Columbia coasts. The whales of each pod line up next to their pod mates, facing the other pod. The pods swim toward each other until they are 50 to 70 feet apart, stop for about 20 seconds, then dive and head for one another again. Next comes the serious business of greeting, where the whales move freely among the crowd and touch other whales as if to say howdy. Whether this greeting ceremony is a universal killer trait isn't known, since it has not been observed anywhere else in the world. On the other hand, killer whale language, a series of clicks and whistles, does seem universal. Like human language, it has its dialects. For example, the well-studied pods of the Pacific Northwestern U.S. and Canadian coasts use a different dialect from pods to the north of them, and each pod uses its dialect in a different way. In fact, individual whales make distinctive sounds that presumably signal other whales who's talking. However, researchers have yet to translate the various sounds into meaningful messages.

The intense scrutiny of killer whale behavior is a sign of the change in human attitudes. Once hunted by fishermen for destroying their nets and their catch, the enigmatic cetaceans have become quite a hit on the Washington coast, where people can pay to adopt one. Whalers' tales of killer packs attacking great whales and tearing out their lips and oil-rich tongues seem less important today in light of findings on their social system and communication skills. Only one attack on a person has been recorded: a surfer off Big Sur, California, who was severely bitten. The killer whale is a complex animal that deserves to be regarded as more than an efficient hunting machine. We can look forward to finding out just how sophisticated it is as research continues.

The Elephant's Marine Cousin

The manatee, a bulbous creature as long on Earth as the first whales, has been swimming around peacefully for all those 45 millions years. Almost hairless, a bit wrinkly, and with a mouth that moves much like the trunk of an elephant—its relative—it snarfs up about 100 pounds of sea plants every day. Manatees are large, tipping any scale that would take them at 800 to 3,500 pounds, and extending 10 to 12.5 feet. Some believe that manatees may have given rise to mermaid legends because of ther docility and their mermaidlike tails.

These friendly blobs live in warm Florida waters—ocean or rivers—and seem to have no real social hierarchy. They spend most of their day eating and playing, and sometimes come up to boats to be petted. Their only enemy is the human motor boat, which has left scars on some of them.

Feats of Animal Migration

As though their lives depended upon it, gray whales swim from Alaska to Mexico, bluefin tuna from the Gulf of Mexico to Norway, green turtles from the Brazilian coast to tiny Ascension Island 1,400 miles out into the Atlantic, and eels from both the North American and European coasts to the depths of the Sargasso Sea. The lives of living things, these and many others, do depend upon their vast traveling, of course, which is why they have evolved with direction-finding skills as elaborate as inner sun compasses, heightened sensitivities to smell, magnetic and electric sensors, and a facility for learning landmarks. But much remains unknown about the miracles of animal migration.

First is the greatest mystery of all: How could migration have evolved to its present forms? It makes, indeed, an efficient, though elongated, geography, allowing animal species to inhabit more than one ecological niche with all the attendant advantages of food and shelter. Through it, they can synchronize their lives as a species, timing their reproductive periods, for example. But these arguments are misleading, focusing as they do on the level of the species, instead of the individual, where evolution appears to first happen with each animal pursuing its own survival. Migration is, of course, adaptive for an individual animal—those that go the wrong way clearly perish, thereby failing to pass on their genes; those that go the right way together find food and a longer life during which to reproduce. But, paradoxically, certain species that long ago began to go the right way—the island 1,400 miles away or across the Sahara at its widest point (where the European barn swallow traverses it)—seem as though they should have perished too. Some migrations are puzzling.

The answer may lie in the shifts of geography itself. When the Atlantic was only a small ocean between splitting continents, when the puzzle pieces of the continents were in different positions or when the climate was molded by a different stage in the swing of the glacial cycle, when an area of land was not overgrazed, a given migratory route may have been essential, even though it now looks exaggerated. In the case of the green turtles who find Ascension Island far out into the Atlantic, genetic tests have been done that seem to rule out continental drift as an explanation, in part because the turtles that go there and those that go elsewhere are not different enough to argue for a long, stable route. But the barn swallow that traverses the Sahara at its widest point has not yet been compared to other groups of its species; perhaps it began this route when the Sahara was green. Major rearrangements of geography within the last few million years have altered climates within the last tens of thousands. And it is plausible that animals are conservative, still following old avenues toward known rewards.

Their techniques for travel—for both long migrations and short journeys—are no less curious. The first, an inner sun compass, is most common among insects but also occurs among fish and amphibians. This compass is linked, especially in bees and ants, to an extraordinarily acute sense of time. Bees fly to buckwheat blossoms, for example, only in the short morning time when they yield nectar. They then extend this sense of timing to a study of the sun, finding their way home to hives in all seasons and in all weather by noting and remembering the direction of the visible sun and of its polarized light. Streaks and zones of this light, invisible to the human eye, are created by wave vibrations at right angles to the direction of the sun, even on cloudy days. Ants can also use the sun to reach their hills again, allowing for its movement and generally assessing it even more flexibly than bees. Other users of time and the sun include birds; locusts; beetles; water-striders; wolf spiders; small sea crustaceans; many fish, some of which migrate back and forth to the edge of their continental shelves or even farther; some frogs and toads; and green turtles. Fish and amphibians also use an extremely sensitive sense of smell to guide them, identifying the smell of their home waters and of moving currents at sea.

Other common and quite extraordinary techniques found in fish, amphibians, insects and other groups are magnetic and electric sensing. Creatures as diverse as dolphins and honeybees, bacteria and tuna, and turtles and homing pigeons have been shown to bear magnetic particles within their bodies. These allow them to be sensitive both to the Earth's magnetic lines of force and to magnetic anomalies along the ocean bottom or across land. The magnetic minerals could also help them to detect natural electrical currents created in the sea as salt water, a conductor of electricity, passes through the lines of the Earth's magnetic field. Eels may use this sense to follow fast currents in migration, much as sharks and rays are known to use it to detect prey.

Mammals, whose brains are more complicated, seem to use the sun and smell less, and landmarks and learning more, in their migrations. Though the striped field mouse relies strongly on its sun compass, the gray whale supplements its by learning the sea route from the older animals (who use echolocation to follow known contours of the sea bottom). Bats also learn landmarks so that they can use their sonar for homing. Mammals have more alternatives to migration too, including hibernation, toughing it out, and nomadism (regular wanderings to follow a food source). The world is a special home to those who can move skillfully across it.

The Earth's First Inhabitants

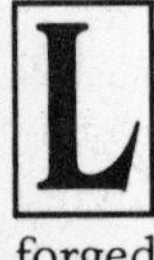

Life began under skies still bright with lightning and thick with volcanic gases, lying under the ultraviolet glare of the young sun. Its precursor molecules were probably forged within the air itself, rained down to seed the sea. A broth

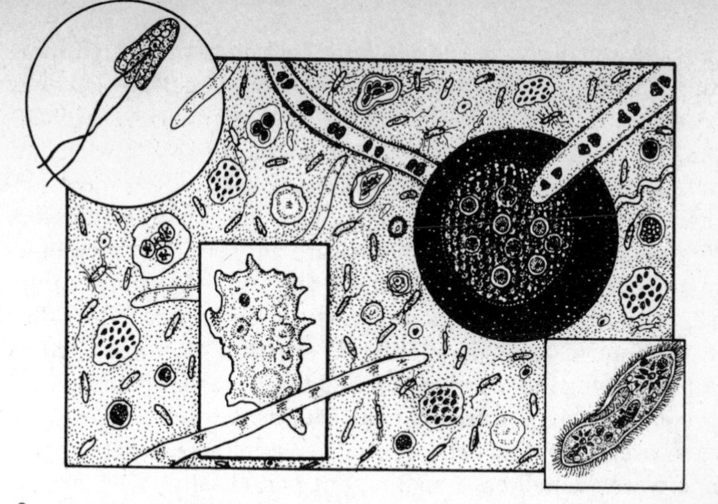

of ammonia, methane, hydrogen, water, small amounts of oxygen, and perhaps carbon dioxide yielded these first molecules, some 4 billion years ago, when the Earth was only about 600 million years old. Several subsequent stages led to other early living things.

Bacteria—the first form of truly independent life—were blended in the oceans some 600 million years after that, a full 3.5 to about 3.4 billion years ago. They were born into still near-boiling oceans, and some forms like them continue to thrive close to volcanic vents deep under the ocean today, converting inorganic chemicals into energy in 480-degree Fahrenheit temperatures, which may well be similar to those in the primitive ocean. Simple, one-celled life forms like bacteria and, soon after, algae ruled the seas at this time and for 3 billion years after it, lapping at the shores of an arid and empty land. Fossilized mats of these algae, the size and thickness of small coffee tables, remain in shallow seas in some areas of the world today, notably Australia. Called stromatolites, the oldest have been dated at close to 3.5 billion years old. Evolving slowly, the early creatures first ate carbon molecules, then learned to spin their food from sunlight through photosynthesis. For this they needed hydrogen, gleaned first from volcanic gases, then from water. When bacteria developed the ability to split water into hydrogen and oxygen, they created the beginnings of our oxygen atmosphere.

But life remained in the sea for a long time, protected there against the sun's ultraviolet radiation, still largely unscreened by an atmosphere. Some 10,000 species of protozoans developed, including the paramecia and amoebae still alive today. These life forms are essentially immortal, splitting into new one-celled clones of themselves without ceasing. Then, well before 600 million years ago, came a creature very similar to the modern volvox: a definite collection of cells, and the first creature on Earth to experience sex and death. It can perhaps be called the first life in any way similar to our own. A jeweled green marble bobbing in the sea, a hundred times smaller than a child's marble, the volvox came to hold within its watery body some five to twelve floating cells, specialized into three distinct types. Some were "daughters," spilling out periodically to begin new life in the more primitive way of the one-celled forms. Others were very like sperm cells in later, higher animals and still others like egg cells; they fused together periodically to launch life in the more advanced, sexual way. Some volvoxes even evolved to produce either eggs or sperm only.

Life continued to proliferate into sponges (large groupings of cells), jellyfish (the first organisms to have a nervous system and muscles), worms, and many other soft sea creatures. Their rule lasted until about 570 million years ago.

At that point, life, as seen in the fossil record, exploded again. Marine animals somehow came to spin calcium into shells. This protection allowed further bursts of evolution. By 400 million years ago, the seas had filled with life. The fishes, on the planet 100 million years before any other vertebrates, still comprise about half of all the planet's vertebrate inhabitants.

Most of the land probably still remained to be colonized, though the very first plants appeared on shore as early as 400 million years ago, descendants of the algae. "Land" at that point was probably only large granite islands here and there, not yet true continental masses. As marine algae still lapped at the splash line, worts and mosses may have moved a bit farther inland. Plants don't leave fossils very easily, but the first true land fossil is of a thin, leafless, branchlike shore plant just a couple of inches tall. The insects followed the first plants quite quickly, the earliest known one being a tiny Devonian bristletail, wingless and similar to today's silverfish, found in a 390-million-year-old stratum; by 360 million years ago, the fossil of a cockroach shows a creature very similar to today's roach. Then, as long ago as 300 million years, ferns were making the world bloom and animals with skeletons were coming into abundance on land.

In the sea, marine organisms had by this time exhaled a great deal of oxygen, creating a high layer of ozone that still protects us, mostly, from the sun's ultraviolet rays. The land plants began to breathe out much more oxygen as well. It was they who had made—and are still making—the air for the first large populations of arthropods—centipedes, millipedes, and mites—and all the animals that followed them into life over the next 230 million years and more.

There are now, on Earth, between about 5 million and 50 million species, quite a span of uncertainty in a world richly beautiful but not completely discovered. All hold the same central molecules of life.

The Well-Protected Plant

Short of swinging a sharp left branch or stinging with the mouths of their flowers, plants are indeed amazingly adept at protecting themselves against insects and other animals. Though some of their weaponry is physical, most of it is chemical. Some 10,000 defensive compounds produced by plants have been discovered already. With a life on the Earth's land of more than 400 million years, plants have had time to evolve an arsenal that includes insecticides, fungicides, antibiotics to fight bacteria, and substances toxic to many animals (including humans), as well as to other plants. The green world is neither passive nor pacifist.

One method of protection is physical. Though blending into the background is not a common strategy, at least among flowering plants—which, after all, need to advertise themselves by color, scent, or other attributes to attract pollinators—a few do use camouflage. One kind of succulent, called the "flowering stone," grows between desert pebbles, and is itself a rocklike gray with a pair of angular leaves. Only in a short but exquisite flowering season is it visible to hungry animals. Many roses, cacti, and berry plants use armor too, of course: a range of prickles, thorns, and spines designed primarily to prevent birds, deer, and other animals from taking a bite.

More abstract physical protection includes forms of mutualism and parasitism. Some trees grow hollow parts to make homes for ants or other stinging insects, which then serve the tree as guards. An acacia tree of Central and South America, for example, harbors biting ants in hollow thorns. True parasitism lies within the plant kingdom too. The Indian paintbrush looks out for itself by "preying" upon the roots of neighboring plants, depriving them of some nutrients but not actually killing them as it grows into their roots. Most mistletoes have no true roots of their own at all; instead a rootlike apparatus taps into the vascular system of the branches of trees and shrubs, drawing water and nutrients. Perhaps the most rapacious plant parasite is the dodder, called the "devil's sewing thread." Its thin, twining stems encircle its host, penetrating the stalk and almost draining the other plant dry.

The chemical arsenal of plants is even more subtle and widespread. One powerful group is thoroughly antiinsect; some are antimammal. Wounded leaves of plants such as potatoes, tomatoes, tobacco, eggplants, clover, alfalfa, soybeans, wheat, rice, corn, and others manufacture proteins that interfere with the digestive enzymes of animals. Plants can also chemically interfere with the hormone-based growth and development of insects, as the bugleweed does to army worms, or literally poison them, as mustard plants do to the black swallowtail butterfly. Once chemical production begins, the protective chemical appears in unwounded leaves too, in order to deter further munching. Other plant-made products that operate against insects include various tannins and lignins; toxicants produced when the plant is thirsty (warding off trouble when the plant is especially vulnerable); protective chemicals called alkaloids, which interfere with the nervous systems of pests; some amino acids that make defective proteins within the eater's own body; good quality cyanide; and other substances that destabilize the cell membranes of the plant's predators. The effects of these toxicants can be long-lasting. Larch budmoths unfortunate enough to attack a larch tree en masse will stimulate the production of less edible needles that then slow the insect population's expansion considerably over a period of about five years. Gypsy moths can be slowed down in similar ways by gray birches, black oaks, and perhaps other trees. In fact, it may not only be temporary weather or short-term climate fluctuation that govern the expansion and contraction of insect populations—it may well be the trees themselves. Trees under attack by insects seem able to "communicate" danger signals to nearby trees as well, stimulating them to produce their changes in leaf chemistry. Willows, sugar maples, red alders, and poplars are among those already known to "talk" in this way. They also seem to be able to coordinate their flowering, perhaps in order to attract more pollinators. And, even within a single tree, different leaves can produce different defenses.

Another kind of chemical protection against insects is found among conifers, which use their sticky resin to trap infesting pests, sometimes entombing them as mummies in amber, the petrified resin. And the odor of the citrus trees' peel is instantly fatal to many insects.

Plants also produce chemical fungicides to protect themselves against various fungi, and antibiotics to protect against bacteria. Potatoes can produce, within several hours, both anti-fungal and antibacterial agents called phytoalexins, which inhibit the growth of microorganisms. More than 100 of these protective antifungal chemicals have now been found all over the plant kingdom, with soybeans mounting their defense against even a billionth of a gram of attacking organism. The reaction is usually localized and quick, occurring within seven or eight hours.

Natural antibiotics have also been isolated in tobacco, sycamore, soybean, and wheat plants. Many plants can act even faster too, sacrificing a few cells very early to the invader, in a way that seems to slow down the spread of the pathogen even before the fungicides and antibiotics are produced.

Chemical protection against other animals is also common. Many mushrooms can poison birds and mammals, with some animal species unaffected by a mushroom that can kill another. Milkweed makes blue jays sick. The sensitive mimosa plant goes droopy and limp when touched, partly to make itself look unappetizing. And many substances in plants are simply bitter enough to deter eaters.

Plants can even chemically protect themselves against other plants that infringe on their territories. The guayule of Mexico secretes cinnamic acid from its roots, a poison to other plants' nearby roots. The black walnut's fruits, when they begin to decompose after falling to the ground, seep a toxin into the soil for the same reason. The sycamore has a natural herbicide in its leaves and fruit balls, which, when they drop to the ground, prevent many other plants from growing under the sycamore. Sagebrush, sunflower, and sorghum are also territorial.

The Edible Library

Plenty of insects and other tiny animals eat books: moths, cockroaches, booklice, silverfish, termites, and bookmites. They love the organic glue in old bindings, the invisible fungi on the damp pages, and the covers, in both old and new recipes. The only way to do these creatures in is to put the books in the deep-freeze (−40 degrees Fahrenheit) for about three days.

Do Humpback Whales Really Sing?

If Hollywood ever gives an Oscar for best performance by a cetacean actor, the humpback whale would be hard to beat. From Japan to Hawaii and from New England to the Caribbean, humpbacks have cavorted their way into the hearts of thousands of scientists and other whale watchers. But while we're handing them an Oscar for their antics, perhaps we should also present them with a Grammy for their peculiar, haunting song, one of the most fascinating features of whale behavior ever discovered.

Divers swim alongside these amiable leviathans without fear of attack, watching a calf glide above its mother or a group of adults feed in great choreographed rushes. Humpbacks put up with closer human contact than other whales; the cows even let people within a few feet of their calves. Humans badly betrayed that trust in the nineteenth century, when whalers harpooned calves and held them hostage to lure their mothers within range. Modern whaling also took a great toll, so much so that the species teetered on the brink of extinction. In 1966, an international treaty saved them by banning their commercial killing, and now they seem to be doing well.

Humpbacks are easy to identify, with their characteristic blows—plumes wide relative to height—and huge flippers. Ranging from one-quarter to one-third the whale's body length, humpback flippers dwarf the flippers of all other cetaceans, including the mighty blue whale. These flippers give the humpback better than average maneuverability underwater and an appearance that suggested its Latin name, which means "big-winged New Englander."

Their tolerance for nosy humans has given humpbacks a reputation for gentleness, but with each other they don't hesitate to play rough. During breeding season males frequently jostle each other, sometimes drawing blood as they scrape their barnacle-encrusted heads and flippers against their rivals. While no one has witnessed humpback mating, cow-calf pairs are often seen in the company of "escort" males as they patrol the warm-water breeding grounds. The escort usually swims a little behind and below the cow, waiting, it is thought, until she is ready to mate with him. If another male tries to take his place, the escort first simply tries to stay closer to the cow. But if the intruder persists, the escort will lunge at him in an all-out effort to keep from being displaced. Females usually give birth every two or three years, but off the Hawaiian coast some have been seen with a new calf every year.

Humpbacks range along the Asian and eastern Pacific coasts and the western Atlantic. They spend about six months of the year feeding in cold Arctic or Antarctic waters, or cool California current waters, where they put on thick layers of blubber to last them through the rest of the year. When winter approaches, they head for tropical mating and calving grounds and the completion of their annual migratory circuit—a round trip of perhaps 6,000 miles. During their stay in the higher latitudes the whales seem to forget the rivalries and fights of the mating season. Groups of males, females, or mixed sex form and reform as the whales freely associate with one another for feeding and socializing. Humpbacks, like all other baleen whales, feed by taking in huge mouthfuls of water and straining them out through the fringed row of hard baleen plates that hang from their upper jaws in place of teeth. Food—mostly inch-long crustaceans called krill—gets trapped in the baleen curtain, licked off, and swallowed.

The humpbacks don't just take whatever sample of water is in front of them, but actively herd prey to their doom. North Pacific whales in particular have been seen circling rapidly under and around a school of fish, all the while releasing a stream of bubbles. After the prey have been surrounded and confused by the bubbles, the whales lunge toward the surface, scooping up the hapless fish in the process. Called bubble netting, this method of fishing makes intriguing whale watching. The first sign is a ring of bubbles, about 10 or 15 meters across, rising through the water. A couple of seconds later a gaping whale mouth breaks the surface, holding up to several hundred pounds of fresh food. Sometimes whales bubble net together, in which case the whale watcher is treated to the spectacle of several huge whale snouts bursting into the air close to each other, as if performing a water ballet.

Of course, it is singing and not dancing that sets humpbacks apart. Singing appears to be the male's mating call, a distinctive series of notes repeated over and over again in the same sequence, for up to half an hour at a time. The singer belts out his song suspended upside-down, up to 20 meters deep, with no other whales around. Recent work indicates that humpbacks even incorporate their own form of rhyme into their songs. Just as certain sounds in our languages rhyme to us, so certain sounds such as clicks and whistles seem to rhyme to them. The more complex the song, the more the humpbacks rhyme, which may help them remember how it goes. At the start of the breeding season all the whales sing the same song, but as the season progresses they gradually modify it. By season's end they are all still crooning a single tune, but it is completely different from the one they began with. The whales quiet down for the summer feeding period, but on arriving back in the tropics they take up the same song they left off with at the end of the last breeding season. It, too, will be revised until it has evolved into a completely new melody.

Humpback songs carry for up to five kilometers under water and sometimes attract another male. In that case, the singer and the other whale swim together for a few minutes, then go their separate ways. Usually, one will renew the song after the parting. Some scientists think the encounter is the intruder's way of interrupting the song before the singer can attract a female. Singing is usually done by dominant males that seem to be advertising their fitness to females with songs, much as a peacock advertises his with feathers.

Other humpback displays are harder to link to specific motives. In a common behavior called breaching, the whales shoot straight out of the water, twist in mid-air, then fall on their backs with a tremendous splash. In lobtailing, the whales slap their tail flukes hard on the surface. Finning, also a frequent display, amounts to lobtailing with the flippers instead of the flukes. A fourth aerial maneuver, in which the whales throw their flukes high in the air as they start to submerge, almost always precedes a deep dive that may last 15 or 20 minutes.

Why they do all these things remains, for now at least, the whales' secret. Rival males competing to escort a cow often breach as part of their aggressive interaction, and escorted cows may do so too. Also, whales apparently sometimes breach in reaction to human activities. For example, humpbacks off the Alaskan coast once breached repeatedly as a cruise ship passed by over a mile away. Scientists recording underwater sounds found that the breachings closely followed noise spurts caused by the ship's engine being revved up.

Since large whales can't be kept in captivity, clues to the meaning of humpback behavior will continue to depend on observations in the open ocean. This lovable whale has much still to teach us, if we listen hard enough.

How Dangerous Are Vampire Bats?

As kindling for the human imagination, the vampire bat has no superior. From Bram Stoker's chilling novel "Dracula" to modern, campy vampire spoof movies, this medium-sized bat has spawned a legend far out of proportion to its size and ferocity. It never haunted the hills of Transylvania,

and was unknown in Europe until the sixteenth century, when reports of "blood-sucking bats" filtered back from New World colonists. Although the vampire doesn't really suck blood, it still inflicts enough damage to make it one of the most respected and feared pests in the Western Hemisphere.

Vampires come in three species, all native to the Americas. Best known is the common vampire, which feeds on the blood of mammals. Its cousin, the hairy-legged vampire, seems to prefer bird blood, while the white-winged vampire appears to like both birds and mammals.

A typical vampire is 7 centimeters long, weighs 30 to 35 kilograms, and has a 40 centimeter wingspan. It looks ordinary enough except that its cheek teeth have been reduced to nubbins, while the incisors and canines have become razor-sharp. The teeth just behind the canines are used to shave fur or feathers from victims in preparation for a bite. The grooved lower lip draws blood by capillary action toward the mouth as the bat laps up blood from the wound with its tongue.

The common vampire usually feeds on large domestic mammals such as cattle, horses, sheep, and goats. The bat approaches at night, while the victim is sleeping, using the fleshy pads on its hands and feet to soften its landing on the victim's body. Usually, areas with thin fur or hair and a generous blood supply, such as the neck, ears, and legs, are chosen for the incision. The bat bites quickly, injecting an anticoagulant into the victim's blood to keep it flowing, and drinks 15 or 20 milliliters if undisturbed. If enough blood is drawn, several bats may feed at one bite, or several may visit in turn. In any event, the anticoagulant can cause serious bleeding if the victim is a young animal or is subject to repeated attacks. A single bat will consume perhaps 7 liters of blood a year; a colony of one hundred bats can drink the blood equivalent of 25 cows or 14,000 hens.

Since blood is mostly water, a bat must drink until it is bloated like a balloon in order to get enough calories and nutrients. But water is heavy, and so the bats begin to urinate almost as soon as they have begun feeding in order to stay light enough to fly away. The vampire's metabolism is also adapted to safely handle the large amounts of iron it absorbs with the blood's hemoglobin.

After a meal, vampires return to their roosts in caves, old buildings, hollow trees, or other sheltered places. They may not venture out again for several days, particularly if the moon is bright. Avoiding the moon may have something to do with evading birds of prey that eat bats.

What vampires fed on before Europeans and their domestic animals invaded the New World isn't clear, nor is how they became so dependent on a blood diet. Perhaps their ancestors devoured parasites such as ticks and blood-sucking insects and had to evolve methods of dealing with large amounts of blood in their food. Then they found ways to bypass the insects and other blood-suckers entirely and became parasites themselves. With the arrival of Europeans and their huge, docile animals, the vampires' food supply multiplied and so did their numbers.

These voracious little bats don't restrict themselves to four-footed animals. People sleeping outdoors with insufficient protection may wake to find themselves marked with the bloody splotches characteristic of the vampire. The big toe is a favorite target, but cheeks, fingers, or any other uncovered part may be attacked. An increase in the human population, along with increased poverty, in South and Central America means more people sleeping outside, and so more bites.

Vampire bites may sound horrifying, but they are not dangerous in themselves. The real peril is from diseases such as tetanus or rabies that the bats may carry. Incidents like the rabies outbreak in Trinidad during the 1930s, which claimed 89 human lives and thousands of cattle and was blamed on rabid vampires, prompted scientists to look for a way to control this bat. Trapping them, napalming their roosts, and other drastic measures did little good, but a bit of chemical trickery produced results.

By closely watching the bats, scientists found that a bat returning from a meal will thoroughly groom itself and several fellow bats. An anticoagulant mixed with petroleum jelly and applied to wound sites on farm animals will soon be smeared over any bat that attacks the animal and carried back to the roost. There, the poison will be spread to several bats as they groom each other. Alternatively, an anticoagulant can be injected into the blood of cattle in such a small dose that it will not harm the cow but will be fatal to a bat that drinks the blood.

Such control methods will not wipe out vampires, but will enable livestock to survive and share their territory. It would be sad if this bat disappeared, for despite its potential for harm, the vampire remains one of the most enduringly fascinating animals on Earth. Just like the legends.

The Avenging Ladybug

Though the honey bee is the insect most helpful to us as a whole, there is another insect who kills pests for us, performing an invaluable service. It is the ladybug—more correctly the ladybird beetle—and it almost constantly consumes the aphids, mealy bugs, and scale insects that destroy our farm and garden crops. Even in their larval stage many species of these bright creatures eat enemies. And one ladybug, as an adult, migrates 50 miles or more down valleys looking for aphids, ingesting 100 to 300 of them in order to replenish its depleted fat stores.

Besides ladybugs, human agriculture is also helped by many other predatory and parasitic insects, including wasps and some flies. But the brass ring goes to the creature so prized by early English farmers that, when they had to burn off their hop vines, they feared for its safety and said, "Ladybug, ladybug, fly away home,/Your house is on fire, and your children will burn." An old favorite nursery rhyme for an old favorite insect.

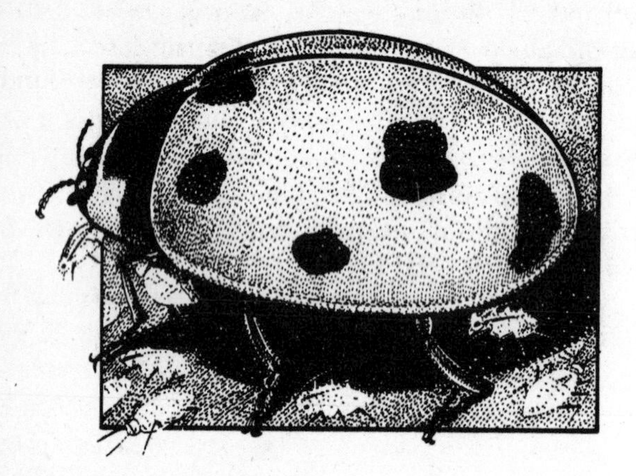

Nature's Mimics

Near the surface of the Amazon, a small fish swims leisurely in search of food. Seeing a leaf floating nearby, it pays no attention and passes close to the stem. Suddenly, the leaf comes alive, grabs the fish and eats it.

The little fish was ambushed by an Amazonian leaf fish, a

creature whose body is flattened from side to side like a leaf. A black line runs along each side like the midrib of the leaf, and a fleshy tab on its lower jaw looks just like the stem. Lying still with its fins drawn in, the leaf fish looks enough like its botanical model to fool most prey that come its way.

Nature abounds with animals and plants that mimic other organisms, often in appearance but sometimes in sound, smell, or other ways. Its deception may bring the mimic protection, reproductive success, or, as with the leaf fish, a meal. Some cases of deceptive coloration or anatomy, such as insects that look like sticks or leaves, are better termed camouflage, for their function is to make the animal blend in with its surroundings and so avoid notice. The list of mimics and camouflage artists is a long one, but a few examples can illustrate the astounding precision with which one species can copy another.

One of the most famous mimics, the viceroy butterfly, copies the brilliant orange and black pattern of the familiar North American monarch butterfly. While viceroys taste good to birds, some monarchs taste bitter and cause birds such as blue jays to vomit, a result of the butterfly spending its larval days feeding on milkweed plants high in distasteful chemicals. When birds encounter monarchs they quickly learn to avoid them, and in the process also learn to avoid the harmless viceroys. Thus the viceroys gain protection from mimicking the monarchs. In Florida, the monarch is rare, but its relative the queen butterfly is abundant and poisonous. There, the viceroy mimics the mahogany coloration of the queen and so again "fools" predators into avoiding it. Sometimes two poisonous butterfly species will have similar patterns. That way, a predator has fewer patterns to learn to avoid and so is likely to make fewer fatal (for the butterfly) mistakes.

Ants, many of which repel predators with their acidic taste, have many mimics, including spiders. In India and Africa, a red spider is often found close to colonies of red weaver ants, and black ants have black spider mimics. The ants are not fooled by the mimicry, and the spiders avoid them, for the red ants especially can be quite vicious. Instead, the spiders seem to fool spider-killing wasps. Black ants rarely wind up in the wasps' nests, and red ones never do. It seems the extreme ferocity of the red ants, who are more than capable of killing wasps, is enough to keep wasps away from their bailiwick.

Wasps have their own mimic in the hoverfly, which looks dangerous but lacks a sting. Similarly, the dronefly looks and buzzes like a honeybee, but can't sting like one. Toads that have been stung trying to eat bees will avoid both insects, and there is evidence that the mimicry of the bee's buzz also helps turn the toads against the dronefly. The robberfly mimics the bumblebee, and so gains protection against toads that have learned to avoid them. The robberfly repays the bumblebee by preying on it, aided by an appearance that seems to help it get close to its unsuspecting prey. Robberflies also eat honeybees, wasps, and other insects.

Moray eels rank high among reef fish that are not to be messed with. One type, dark-colored with little white or bluish dots, has a mimic that is anything but fierce; a fish that hides its head in crevices in the reef. It leaves its tail sticking out, but has little to fear from predators because the tail is marked almost exactly like the moray's head, complete with an eyespot (actually just coloration, not an eye) that mimics the moray's eye.

Another reef fish, the cleaner, lives off parasites gleaned from other fish. The cleaner occupies a "cleaning station" and signals to its "customers" that it is open for business by performing a special dance. Sometimes customers even queue up for its services, which include nibbling inside their gills and mouth to remove parasites and dead skin. But lurking near the cleaner's territory may be a blenny that looks much like it and lures the cleaner's customers with the same dance. Unlike the fish it resembles, however, the blenny bites the customers when they're not looking.

Some of the best mimics are plants. Many orchids trick male insects into pollinating them by mimicking the bodies of female insects. The South African starfish flower uses not only sight but smell to fool the flies that pollinate it. Colored like rotting meat, the malodorous flower attracts female flies, who lay their eggs in the big blossoms. The maggots die from lack of food, but the females' visits do the job of pollination. The bolas spiders, found in the Western Hemisphere, also use smell in mimicry.

These spiders don't build webs; instead, they hang by a thread from plants and hurl a sticky silk ball at passing insects. But the insects don't just happen by. The spiders produce a scent that mimics the sex-attractant-laden scent of female moths, and male moths follow the spider scent upwind to their doom.

The camouflage specialists employ a wide variety of strategies. Chameleons can turn green, yellow, brown, or a combination of colors in a couple of minutes. Ocean dwellers such as sole and octopus perform similar feats. Stick insects look just like twigs or dead leaves, and an Amazonian katydid's green wings copy the look of a leaf right down to brown "rot" spots tinged with yellow and a realistic pattern of veins. A grasshopper of the Namib Desert changes colors to match its environment and orients itself so as to cast the smallest shadow. From fawns to zebras, animals use spots and stripes to break up their outlines and appear less conspicuous. Seasonal changes are also quite common; the snowshoe rabbit and the weasel change to white coats for winter, and at least two birds, the snow bunting and ptarmigan, do likewise with their feathers.

Another bit of trickery, the bogus eyespot, is employed by several animals besides the moray eel mimic, including caterpillars, butterflies, at lease one South American frog, and some marine fish. Some tropical fish sport a large dark spot, resembling an eye, on their rear flank. Meanwhile, the true eye is concealed by a vertical stripe on the head. A predator that took the spot for an eye would expect the fish to flee in the wrong direction, perhaps giving it a life-saving head start. On the back of the frog, the eyespots, combined with dark thighs resembling a mouth, present a crude "face" to a predator approaching from behind. Some tropical caterpillars can swell their thoraxes, revealing huge eyespots and taking on the appearance of a snake. Also, in experiments with captive birds, the birds behaved as if genuinely frightened of some butterflies' eyespots—at least until they got used to the insects. But if a person rubbed the eyespots off, the butterflies were eaten.

It is said that Mother Nature can't be fooled, but some of her creatures surely can be. Their mistakes allow other animals a chance to survive so that their descendants may refine their incredible fakery even more.

The Ice Worm

This 3.4 inch long reddish brown/black worm lives year-round in Alaskan and Pacific Northwest glaciers. It has its own internal antifreeze and probably eats spores, bacteria, algae, and pollen, but no one is absolutely sure.

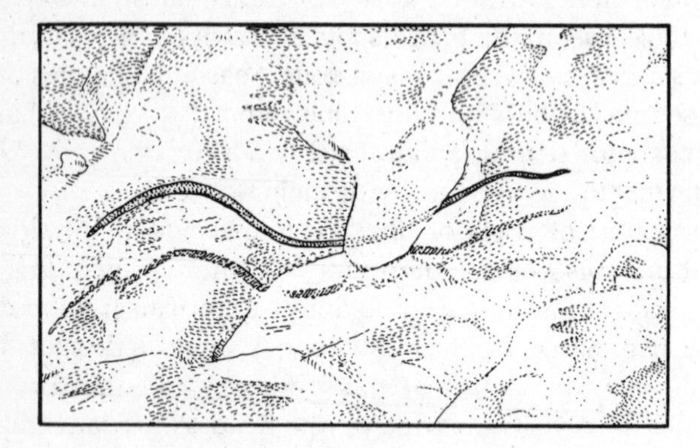

Making a Beeline

The flutter of honey bee to flower, one to sip nectar and pack up pollen, the other to be made fertile, is a dance that probably began when flowers first appeared on Earth about 110 million years ago. Ever since, the two have been engaged in an exquisite evolutionary pas de deux, choreographed by the bee's vision, memory, and "waggle dance" for its fellows, as well as by the flower's timing and other adaptations. The bee has come to pollinate about 80 percent of all flowering plants, at least 100,000 species.

It finds the flowers with its compound eyes. Arrayed in hundreds of tiny lenses, these eyes see the world in flickers, angles of light, and a riot of colors, some invisible to us. Even on a cloudy day, a bee can discern the sun's angle and use it to find the way from hive to flower to hive again. This is a geometric and adept forager.

Once close to a flower, the bee uses both its sense of smell and color vision and can find up to 500 blooms of the same species in one day. Though its eyes are most sensitive to ultraviolet, blue, and green colors, it prefers violet and blue, associating the pure ultraviolet with sky radiation instead. The bee can, however, distinguish all of our colors except for red. Some flowers look different to a bee than they do to us, as its eyes can see far into the ultraviolet range, revealing some unusual markings.

The bee's memory and ability to learn are quite extraordinary, considering that the insect's brain spans only about one cubic millimeter. After a five-month absence from an area, it will fly straight to an old flowering food or water source. And the insects can find such a place immediately, even when food is moved—to try to fool it—several times by a researcher, provided that it is moved the same distance farther or the same angle away from a central point every time.

Once new food is found the bee returns to the hive, on wings that beat 160 to 220 times a second. There it will share the good news in a dance. In this "waggle dance," a series of figure eights, the bee moves with a vivacity related to the amount of food being "discussed." Straight runs up the side of the hive, within the continuing figure-eight waggles, show the direction of this cache of flowers in relation to the sun. For example, a 40-degree angle to the left means that the flowers lie 40 degrees to the left of the sun's position. The type of food source found is conveyed by a nectar or pollen sample that the bee has brought back.

Throughout the long, sunny season of flowers and waggle dances, the bee is busy at home too. First, the hive's structure is attended to; its empty waxen cells are built by the younger worker bees, using flakes of a fatty substance from their abdominal glands, formed to the proper consistency by chewing. The hexagonal cells are assembled in small clusters by bees working from different directions in small teams. The smallest of the three sizes is generally used for storing the flowers' pollen and honey, as well as for hatching new female worker bees. The medium-sized cells are for rearing the male drone bees, and the largest are for queens. A queen lays up to two thousand eggs a day. As the eggs hatch into larvae, they are all fed "royal jelly" (a white paste full of hormones secreted by the workers) for three days. Only after that is a single larva chosen to be the next queen and fed royal jelly every day. The rest, future workers and drones, are fed "worker jelly." The drones are slated for a brief life of trying to mate with the new queen, then death. The workers not only build, but feed the queen and the young and make the honey too. They are the vast majority in a hive that may hold 35,000 to 55,000 bees.

All are now living, and will be all winter, on the flowers' nectar, cultured into honey by salivary enzymes. The honey is stored in the hive cells until it has evaporated to the right consistency, and then it is sealed off.

In exchange for their sweet nectar, the flowers are cross-pollinated, which constitutes their mating. In sipping nectar, the bees sweep pollen from male to female flower and from male to female parts of the same flower, insuring the next generation

and adding to the plants' vigor. To aid in this exchange, some flowers have set up obstacle courses, pummeling the bee with pollen before it can leave. Broom flowers even shoot out pollen from sealed petal capsules when the bee lands. And the bucket orchid of Central America has a nectar so intoxicating that the bee stumbles, then slips into a tiny bucket. To get out again, it must climb up a spout in which it is showered with pollen. Only the grass crops of the world are not dependent upon this busy little insect. Everything else, from almond and apple blossoms to tiger lilies and zucchini flowers, needs the bee.

The link of bee to flower has been a powerful force in their mutual evolution. It has taken untold generations of bees to perfect the search and the dance. But the flowers have adapted too. The joyous seasonal pattern of flowering—buttercups opening in early spring, clover warming in the early summer sun, asters blooming in the fall—is the ultimate result of careful timing by individual flowering plant species. Each flower still alive today has discerned the time when the bees would attend to it. The shared evolution is fine-tuned during a single day as well, with bees visiting morning-opening flowers in the morning and afternoon-opening flowers in the afternoon.

The Revolving Kinkajou

As odd as its name, the kinkajou is one of the strangest mammals in South America. The nine-pound arboreal artist looks like a monkey and was almost classified with the lemurs. Pet shops often call it the "honey bear" because of its appetite for honey, its honey-hued fur, and its stubby, bear-like forepaws. However, naturalists say that it is neither: The kinkajou belongs to the raccoon family.

Like raccoons, kinkajous will eat almost anything they can get their paws on. Fruits are favored, but insects also form a staple in their diet. Excellent climbers capable of descending tree trunks head-first, they hold the distinction of being the only New World carnivore with a prehensile tail. But their most remarkable adaptation to arboreal life is their feet. Blessed with swivel ankles, a kinkajou can hang from a branch by hind feet and tail, turning one or both feet 180 degrees at will—a feat any ballet dancer would envy. Though sometimes kept as a pet, the kinkajou's nocturnal habits—not to mention propensity for climbing—require some adjustments by the owners.

Afterword

These pages tell the stories of remarkable creatures around the world. Many more are known, and countless others await human discovery. But inevitably, a large number will become extinct, some before we even learn of their existence. They will fall victim to the plague of habitat destruction now being inflicted by human expansion in nearly every corner of the globe. The burning of the Amazon rain forest presents only the most spectacular example of a species-rich, relatively unexplored area being swallowed up by human exploitation. What medicinal plants, what helpful garden insects dwell there, will never be fully known if the losses continue. Elsewhere, well-known species also face threats: the panda in China, gorillas in Africa, the orangutan in Borneo. Closer to home, we have only to look at the fate of the passenger pigeon and the ivory bill woodpecker to see that the destruction of species isn't confined to developing countries. Every person who builds a second home on a pristine lake or in a secluded area of woods, or who invests in urban-sprawl development, is part of the same global pattern of encroachment that displaces wildlife and decreases the wild spaces our own species needs for its survival. The battle, however, is not yet lost. Individuals and organizations are working around the clock to bring responsible development, which must include family planning, to growing nations and to save the most vulnerable wildernesses, whether in Antarctica or Arizona. For instance, the Nature Conservancy buys up land and keeps it forever wild; the Sierra Club fights projects such as oil drilling on Alaska's North Slope; the World Wildlife Fund tries to save the exquisite flora and fauna of Madagascar; and other groups pursue likewise worthy goals. If you would like to become part of the solution to the world's environmental problems, the easiest way is to join an organization dedicated to the cause. Here are names and addresses of a few:

National Audubon Society
Membership Data Center
PO Box 2666
Boulder, CO 80322

National Wildlife Federation
8925 Leesburg Pike
Vienna, VA 22184-0001

Nature Conservancy
1815 N. Lynn St.
Arlington, VA 22209

Sierra Club
730 Polk St.
San Francisco, CA 94109

World Wildlife Fund
1250 24th St. NW
Washington, DC 20037

Pioneering conservationist Aldo Leopold said that there are some who can live without wild things, and some who cannot. He meant that some of us don't care about the wilderness, and some of us do. The fate of the Earth depends on the actions of those who do.

—D.M.

Bibliography

Many issues of the following journals and magazines contain articles on the strange and wonderful beings that inhabit the planet:

Annals of the Entomological Society of America
Audubon
Australian Journal of Zoology
Discover
Earthwatch
Harvard Magazine
International Wildlife
Mosaic
National Geographic
National Geographic WORLD
National Wildlife
Natural History
New York Times
Omni
Science
Science '80–'85
Science Digest
Science News
Scientific American
Smithsonian
Smithsonian Research Reports

The following books provide further information about some of the creatures and habitats discussed in this book and other creatures as well:

Baerg, William J. 1958. *The Tarantula.* Lawrence, Kans.: University of Kansas Press.

Barnes, Robert D. 1968. *Invertebrate Zoology.* Philadelphia: W. B. Saunders Co.

Bates, Marston, and the editors of *Life.* 1964. *The Land and Wildlife of South America.* New York: Time, Inc.

Bent, Arthur Cleveland. 1946. *Life Histories of North American Diving Birds.* New York: Dodd, Mead & Co.

Breland, Osmond P. 1948. *Animal Facts and Fallacies.* Harper & Brothers Publishers.

Burgess, Robert F. 1982. *Secret Languages of the Sea.* New York: Dodd, Mead & Co.

Carr, Archie F., and the editors of *Life.* 1963. *The Reptiles.* New York: Time, Inc.

Carthy, John. 1974. *Animal Camouflage.* New York: McGraw-Hill Book Co.

Clausen, Curtis P. 1940. *Entomophagous Insects.* New York: McGraw-Hill Book Co.

De Joode, Ton, and Stolk, Anthonie. 1982. *The Backyard Bestiary.* New York: Alfred A. Knopf, Inc.

Durrell, Gerald. 1984. *Amateur Naturalist.* New York: Alfred A. Knopf.

The editors of Reader's Digest. 1984. *ABC's of Nature: A Family Answer Book.* Pleasantville, N.Y.: The Reader's Digest Association, Inc.

Falla, Sir Robert Alexander, Sibson, R. B., and Turbott, E. G. 1979. *The New Guide to the Birds of New Zealand and Outlying Islands.* Auckland: Collins.

Farb, Peter, and the editors of Time-Life Books. 1967. *The Forest.* New York: Time, Inc.

Fenton, M. Brock. 1978. *Just Bats.* Toronto: University of Toronto Press.

Gordon, Bernard Ludwig. 1977. *The Secret Lives of Fishes.* New York: Grosset and Dunlap.

Heinrich, Bernd. 1989. *Ravens in Winter.* New York: Simon & Schuster, Summit Books.

Henry, S. Mark, ed. 1966. *Symbiosis. Vol. 1.* New York: Academic Press, Inc.

Herrero, Stephen. 1985. *Bear Attacks: Their Causes and Avoidance.* New York: Nick Lyons Books.

Hill, John Edwards, and Smith, James D. 1984. *Bats: A Natural History.* Austin: University of Texas Press.

Howard, Everett S. 1981. *Animal Marvels.* New York: Doubleday.

Keeton, William T. 1967. *Biological Science.* New York: W. W. Norton & Co. Inc.

Konner, Melvin. 1982. *The Tangled Wing: Biological Constraints on the Human Spirit.* New York: Holt, Rinehart and Winston, Inc.

Krutch, Joseph Wood. 1957. *The Great Chain of Life.* Boston: Houghton Mifflin Co.

Longgood, William F. 1985. *The Queen Must Die.* New York: W. W. Norton & Co. Inc.

Macdonald, David, ed. 1984. *The Encyclopedia of Mammals.* New York: Commerce Clearing House, Facts on File Publications.

Marshall, N. B. 1965. *The Life of Fishes.* London: Weidenfeld and Nicolson.

McClane, A. J., ed. 1965. *McClane's Standard Fishing Encyclopedia and International Angling Guide.* New York: Holt, Rinehart & Winston, Inc.

McCormick, Jack. 1966. *The Life of the Forest.* New York: McGraw-Hill Book Co.

Medawar, P. B., and Medawar, J. S. 1984. *Aristotle to Zoos.* London: Weidenfeld and Nicolson.

Moon, Geoff, and Lockley, Ronald. 1982. *New Zealand's Birds.* Auckland: Heinemann Publishers.

Moss, Cynthia. 1988. *Elephant Memories.* New York: William Morrow & Co., Inc.

Nelson, Bryan. 1979. *Seabirds. Their Biology and Ecology.* New York: A & W Publishers, Inc.

Patent, Dorothy Hinshaw. 1978. *Animal and Plant Mimicry.* New York: Holiday House, Inc.

Peterson, Roger Tory, and the editors of Time-Life Books. 1978. *The Birds. 2nd ed.* Alexandria: Time-Life Books, Inc.

Pough, Richard H. 1951. *Audubon Water Bird Guide.* Garden City, N.Y.: Doubleday.

Reader's Digest. 1986. *Sharks: Silent Hunters of the Deep.* New York: Reader's Digest Services Pty. Ltd.

Rickett, Harold William. 1966. *Wild Flowers of the United States.* New York: McGraw-Hill.

Ridgway, Sam H., ed. 1972. *Mammals of the Sea: Biology and Medicine.* Springfield, Ill.: Charles C. Thomas, Publisher.

Schober, Wilfried. 1984. *The Lives of Bats.* New York: Arco Publishing, Inc.

Silverstein, Alvin and Virginia B. 1968. *Unusual Partners: Symbiosis in the Living World.* New York: McGraw-Hill.

Stewart, Darryl. 1984. *The North American Animal Almanac.* New York: Stewart, Tabori & Chang.

Swain, Roger B. 1983. *Field Days: Journal of an Itinerant Biologist.* New York: Charles Scribner's Sons.

Taverner, Percy Algernon. 1939. *Canadian Water Birds. Game Birds: Birds of Prey; A Pocket Field Guide.* Philadelphia: David McKay Co., Inc.

Trager, William. 1970. *Symbiosis (Selected Topics in Modern Biology).* New York: Van Nostrand Reinhold.

Walker, Ernest Pillsbury *et al.* 1975. *Mammals of the World. 3rd ed.* Baltimore: Johns Hopkins University Press.

Watson, George E. 1975. *Birds of the Antarctic and Sub-Antarctic.* Washington: American Geophysical Union.

Index